Density Functional Theory Calculations

Edited by Sergio Ricardo De Lazaro,
Luis Henrique Da Silveira Lacerda
and Renan Augusto Pontes Ribeiro

Published in London, United Kingdom

IntechOpen

Supporting open minds since 2005

Density Functional Theory Calculations
http://dx.doi.org/10.5772/intechopen.78820
Edited by Sergio Ricardo De Lazaro, Luis Henrique Da Silveira Lacerda and Renan Augusto Pontes Ribeiro

Contributors
Daniel Glossman-Mitnik, Norma Flores-Holguín, Juan Frau, Muraleedharan Karuvanthodi, Rakesh Kumar, Sangeeta Obrai, Guocai Tian, Weizhong Zhou, Alexander D'Yachenko, Ivan Mitropolsky, Sergio Ricardo De Lazaro, Safna Hussan Kodakkat Parambil, Hisana Asharaf Thozhuvana Parambil, Shahina Parammal Hamza, Anjali Thirumangalath Parameswaran, Mohamed Shahin Thayyil

First published in London, United Kingdom, 2021 by IntechOpen
IntechOpen is the global imprint of INTECHOPEN LIMITED, registered in England and Wales, registration number: 11086078, 5 Princes Gate Court, London, SW7 2QJ, United Kingdom
Printed in Croatia

British Library Cataloguing-in-Publication Data
A catalogue record for this book is available from the British Library

Additional hard and PDF copies can be obtained from orders@intechopen.com

Density Functional Theory Calculations
Edited by Sergio Ricardo De Lazaro, Luis Henrique Da Silveira Lacerda and Renan Augusto Pontes Ribeiro
p. cm.
Print ISBN 978-1-83881-082-5
Online ISBN 978-1-83881-083-2
eBook (PDF) ISBN 978-1-83881-084-9

For EU product safety concerns:
IN TECH d.o.o., Prolaz Marije Krucifikse Kozulić 3, 51000 Rijeka, Croatia, info@intechopen.com or visit our website at intechopen.com.

Meet the editors

Prof. Sergio Ricardo de Lazaro is married, and he has two daughters living in Ponta Grossa, Paraná. Currently, he is a Full Professor in the Department of Chemistry of the State University of Ponta Grossa (UEPG). He has his Ph.D. degree in Chemistry and has published manuscripts with researchers from other countries. His development of studies in Material Chemistry or Molecules applies the Density Functional Theory (DFT). In scientific journals, his recognized documents internationally contribute to environmental, energy, and potential drugs from photocatalysis, charge transfer, and molecular reactivity. His research interests include theoretical analysis of the structural, electronic, magnetic, and morphological properties proposed paths to clarify chemical mechanisms associated with massive problems. He acts as a reviewer of manuscripts in international scientific journals.

Dr. Luis Henrique da Silveira Lacerda is married and lives in Campinas, São Paulo. He completed his Chemistry Ph.D. at the State University of Ponta Grossa (Brazil). Currently, he is a postdoctoral researcher at the State University of Campinas (Brazil). Over the last years, Dr. Luis Henrique da Silveira Lacerda published several works offering exciting insights on materials properties and a new understanding of some materials features at the molecular level. His research experience includes DFT investigations of multiferroic materials, crystalline structure, phase transition, excited states, heterostructures, morphology, and advanced materials properties such as magnetic, ferroelectric, elastic, electronic, photocatalytic, thermoelectric, optical, magneto-optical, and others. His current research interests include DFT investigation of photocatalytic, chemical reactivity, and properties morphological dependence.

Prof. Dr. Renan Augusto Pontes Ribeiro is married and lives in Divinópolis, Minas Gerais. Nowadays, he is a Professor in the Department of Chemistry and Physics of the State University of Minas Gerais (UEMG). With a Bachelor's, Master's, and Ph.D. degree in Chemistry, Prof. Renan Ribeiro is a computational materials scientist with recognized publications through collaboration with research centers in other countries. He has experience with the application of Density Functional Theory (DFT) on the design of multifunctional materials in silico, contributing to fields like environmental, energy, photocatalysis, charge transfer, magnetism, and molecular reactivity.

Contents

Preface

The wave function theory based on the Schrödinger equation was first approximately solved by Hartree-Fock-Roothan (HFR) formalism, initiating a wonderful time of computational chemistry. The possibility of modeling and simulating the geometry, electronic density, and other properties in the quantum level for molecules is today, one of the essential areas of natural science. In this new area, the variational principle's resolution is the central question to reach the energies' electronic levels associated with a set of particles; such an approach was very successful for modeling for atoms to molecules. The HFR approach is a numeric approximation method developed to calculate a set of equations similar to the Schrödinger equation for many bodies, for example, the helium atom, the hydrogen molecule, and others. The evolution of the HFR approach was impressive. Its expansion was fast. Many scientists used the method to create the fundamental details of the molecular concepts and properties simulated today. Hartree proposes a modified wave function from the monoelectronic wave functions product or the self-consistent field (SCF). This approximation was essential, but not precise, to create a numerical path to calculate approximately the many body system's wave functions. The consequence was the development of the canonical molecular orbitals with many problems and high potential. Fock contributed significantly to building the quantum operator associated with Hartrees' wave function denominated Fock operator. Now, the Fock's operator would act on the Hartree's wave function extracting quantum information or properties; more importantly, from the monoelectronic wave function product and Focks' operator, the basis for the numerical implementation of the approach has been created.

Meanwhile, Hartree's' wave function is not determined analytically. It is a numerical approach; thus, it always needs to be built. Roothan used the Linear Combination of mathematical functions to create the Linear Combination of Atomic Orbitals (LCAO). This implementation was essential. It increases the precision of the SCF cycle through basis set functions. There is no unique mathematical form for the basis sets since the mathematical equation chooses to build the LCAO used in the SCF satisfies the Fock operator; it is a basis set valid. Then, the simulation of a more complex system was needed to obtain quantum properties.

The Density Functional Theory (DFT) is an improved quantum theory in recent times. However, such an idea was a development from 1960 by Hohenberg-Khon-Sham. Contrary to wave function or many bodies theory, the scientists create a new quantum approach to simulate many-body systems changing the wave function approach using the electronic density (r) concept. The evolution of the DFT showed that the operators have potential, and the monoelectronic formalism is equal to that found in the HFR approach; however, for the bi-electronic interactions in electron-electron repulsion, there was a significant change. In the HFR approach, the correlation energy is not available, resulting in lost information; in the DFT, the correlation energy is exact. It can be calculated from a derivative between energy and the electronic density creating a function. In a sense, since the functional creation, the DFT has three dimensions (x, y, z) for calculations; while the HFR approach has four dimensions (x, y, z, spin). Indeed, the find of the analytical form for the functional is the goal in the DFT. Then, from the last 60 years, scientists have researched numerous functional methods to solve such a mathematical problem. Initially, the functional was described to understand the functional's contribution to the molecular results. The Local Density Approximation (LDA) and Local Spin Density Approximation (LSDA) are examples. Very quickly, the researchers noted this task was not simple. A simple form to describe the functional acted in a path less efficient for molecules with a stronger correlation effect; and, the molecular geometry and electronic properties were poorly calculated. The results were hugely improved from the Generalized Gradient Analytical (GGA) introduction in functional formalism, showing the importance of describing the electronic correlation from an electronic density distribution. A pure functional description was initiated; the foremost functional was the Perdew-Burke-Ernzerhof (PBE).

Meanwhile, functional complexes were created. An example of this new implementation was the PBE0 functional called a hybrid functional. After the hybrid functionals were introduced, the parameterized hybrid functional describes the functional by factors from experimental data to correct the GGA approach. In truth, the great challenge forced by the search of the final functional drove a huge scientific effort for a significant functional until recent days.

In this book, the chapters show research in a broad range of DFT applications. The first chapter is dedicated to one of the essential topics in Computational Chemistry: chemical reactivity. The analysis of the frontier molecular orbitals (FOM), more specifically, the high occupied molecular orbital (HOMO) and low unoccupied molecular orbital (LUMO), indicates molecular properties beneficial to support the chemical reactivity of compounds. The investigations on metallic complexes are very useful because of the molecular geometry, bonding, and antibonding molecular orbitals, charge transfer, and other molecular features. Ionic liquids have a great property of stability concerning significant electric potential variation. The second chapter describes Koopmans' theorem and Fukui index on three liquid ionic families discussing each family, an example of this new technological area's computational concepts. Chemoinformatic or in silico are words to indicate computational simulations in discrete or multiscale levels often applied to biological systems to investigate interactions, and molecular structures are the goal of these calculations. This book then presents one chapter to show an essential tool that can be worked in drug and bioactivity themes. In the same direction, one chapter is on molecular docking studies of the antioxidant effect of the Coumarin compound. The DFT is deeply connected to mathematical formalism; thus, it can be expanded to other areas. The chapter on the non-equilibrium equation is a sample of this expansion, applying the DFT to create a representative model associated with non-equilibrium states. In this case, the proposed model investigated heavy atoms and pions.

Welcome and have a good read!

Sérgio Ricardo de Lázaro
State University of Ponta Grossa,
Department of Chemistry,
Uvaranas, Ponta Grossa, Brazil

Renan Augusto Pontes Ribeiro
State University of Minas Gerais,
Department of Chemistry,
Divinópolis, Brazil

Luis Henrique da Silveira Lacerda
State University of Campinas,
Institute of Chemistry,
Campinas, Brazil

Section 1

Introduction

Chapter 1

Introductory Chapter: A Brief Mention for High-Pressure in Oxides from DFT Simulations

Sérgio Ricardo de Lázaro

1. Introduction

High pressure simulations are employed in materials because of the resistance and advanced properties. The material modifications from high pressures are essential to reach new crystalline structures. In geophysics, many high pressure processes happen inside Earth. Diamonds are a clear example on the matter transformations under high pressures. In this case, the diamond is the natural material more resistance already found; several studies were performed to understand how carbon in graphite form can be changed to diamond and the result was a synthesis procedure based on high pressure in a short time. The initial studies using pressure were performed in gas phase. The definition of the pressure is clearly connected to how a surface is modified from an applied force. All gaseous kinectic theory has the pressure as a central point to understand the atomic and molecular movement in macroscale. In solids investigations, the pressure is the property that deform or change the surface of the material. The high material resistance regarding to gas, the deformations were classified as total and partially elastic or inelastic showing a point of view on reversible and irreversible processes. Such concepts origin the idea on malleability of the solids since metals to plastic. Amorphous and crystalline solids have similar resistance; however, the cleavage of each one is dominated from the molecular ordering. The high pressure finds to investigate the phase transitions in crystalline solids without to reach the crystal's cleavage. The compression modulus theory are the scientific approaches more developed to understand solid deformations from applied external pressure. Since one-dimensional (1D) deformation denominated as Young Modulus, and, posteriorly, Shear (2D) and Bulk Modulus (3D); it is knowledge that Equation of States (EOS) are associated to the energy x volume curve to calculate Bulk Modulus.

From coming Quantum Mechanic, more specifically, the Computational Chemistry in solid state; the making of the energy x volume curve becomes a possibility to calculate the Bulk Modulus from only theoretical proceedings. In this terms, the Density Functional Theory (DFT) has been broadly applied to predict and to investigate material's compressilibity from chemical substitutions or new crystalline structures. Such new approach reaches to a new level for research in Chemistry of Materials area.

2. Short points in pressure

The pressure play an important role in chemical and physical systems. It is defined as force on area (1) showing as an external force acts on substances or materials in solids, liquid, and gas states. Automatically, the pressure is a physical

IntechOpen

property in molecular level only associated to the molecular collisions from a perception to push or to pull molecules causing molecular deformation. The association of the displacement molecular and classical physics from Newtonian mechanism created the gas kinetic theory. This essential approach allowed advances on how the pressure acts on the inside or the outside of a physical system.

$$p = \frac{F}{A} \tag{1}$$

The matter absence drives to pressure absence, there is not pressure without atoms or molecules. Explosions and vacuum are directly connected with the pressure. A recent demonstration was the explosion in the Beirut's port, the energy and gas expansion from the chemical reactions increased hugely the local pressure in a short time crushing lives and buildings. The vacuum is the concept on the absence of matter or the null pressure. This concept was already understood as a breach in the nature, the absence of matter were something unnatural. However, previous experiments revealed that the vacuum is part of nature or the nature does not refuse the vacuum. Today, the vacuum is connected to very low pressure, where same an isolated molecule in a closed vessel can create pressure from the collision with the vessel's wall. Then, the intensity of the pressure is dependent of the molecular motion and amount of molecules.

The pressure and the Gibb's energy variation (ΔG) are connected. The non-expansion work or the spontaneity of a chemical reaction involving gas phase is dependent from pressure. Such concept is expanded for the liquid phase. Raoul's and Henry's Laws investigated the ΔG deviations from perfect behavior how the influence of the long-rang molecular interactions from vapor pressure changes. In solid phase, the vapor pressure is low; however, the pressure can be applied directly on the solid structure. Crystalline and amorphous structures are susceptible to pressure creating solid deformation or crash. In particular, for crystalline structures in deformation process, an external applied pressure changes the Gibb's energy variation causing a phase transformation for other phases. The modifications in crystalline material are connected to spatial groups; it is very interesting because the applied pressure carry out work on the crystalline structure deforming the chemical bonds from "push" the atoms or molecules. This deformation process from pressure is directional, i. e., it dependent of the spatial direction from the pressure is applied. The high ordering implicit to crystalline solids is showed as the Miller indices (h, k, l) connected to crystalline spatial group. Then, the applied pressure can be performed in the three directions of space simultaneously denominated as isostatic pressure; in two directions (bi-axial deformation) or one direction (uni-axial deformation).

Thomas Young, a scientist of nineteen century, investigated the deformation in wires or uniaxial direction creating the Young's Modulus [1]. This study proposes the description of the resistance of a material against the stretch or uniaxial deformation from an external applied pressure or strain. The formalism reached by Young is used until today to measure resistance in several materials as metals, plastic, and composites in automatic devices capable to quantify an applied strength in micro- or millinewtons dimensions. The Young's Modulus, **Figure 1a**, shows in macroscopic dimension, the microscopic forces involved in the continuous resistance inside of the material. In the molecular level, the chemical bonds are on an intense strain to break them. Therefore, the greater the bond strength the greater the material resistance to strain. The bi-axial deformations are knowledge as Shear Modulus [2], **Figure 1b**. In this kind of deformation, the pressure is applied orthogonality on a surface or a bi-axial stretch in x, y-coordinates creating a bidimensional strain on the material. The applied pressure is distributed from strain gradient on all surface

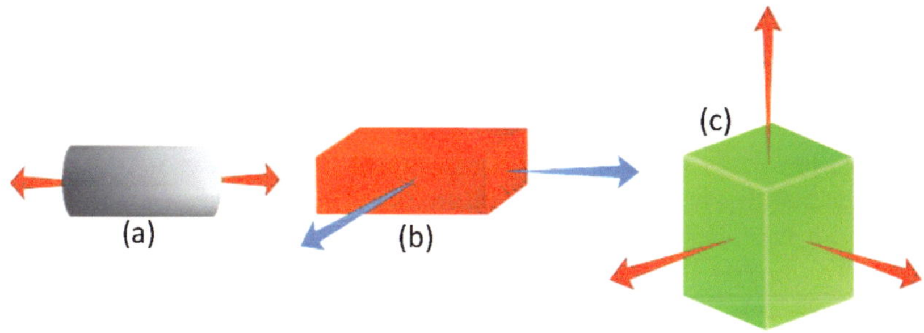

Figure 1.
Schematic representation for (a) Youngu's modulus, (b) shear modulus, and (c) bulk modulus.

or a stress point. Applications are performed in metallic sheet, thin films, and other. The Bulk Modulus [3], **Figure 1c**, is a formalism for three dimensions or volumetric deformation. Such approach is an isostatic procedure; a general experimental procedure is reached from a hydrostatic pressure from hydraulic press. The limitations are the applied pressure by hydraulic press and the resistant container. In general, a sample is insulated from aqueous environment inside of resistant container. Posteriorly, the container is filled with distilled water and closed. The water into the resistant container is compressed applying a isostatic pressure on the sample.

The Bulk Modulus can be positive or negative. The positive value is interpreted as the increase of pressure on the solid causing the decrease of volume; while, if the solid volume increases from pressure increase, the Bulk Modulus value is negative and associated to a thermodynamic instability because of the Le Chatelier's Principle break. The volume always decrease from pressure increase. Then, the Bulk Modulus value is connected to an indication on the energetic instability of a crystalline solid. The Eq. (2) shows the connection among pressure (p), initial volume (V_i), final volume (V_f), and Bulk Modulus (B):

$$p = \frac{B\left(V_f - V_i\right)}{V_f} \tag{2}$$

It is an important comment that for all kind of deformations, the thermodynamic ways are classified as reversible or irreversible processes. In reversible process, the material has a partial or total elastic behavior to reverse the applied strain. In the irreversible process, the applied strain causes a permanent deformation or fracture of the crystalline structure.

From a microscopic point of view, the Bulk Modulus can be calculated from Equations of State (EOS). There is not a general EOS formalism to describe all modifications on crystalline structures. However, various EOS were discovered to represent the possible phase transitions under applied pressure. These EOS are applied in solids with high Bulk Modulus, such scientific theme is called as high pressure investigations. In large scale, the Birch-Murnaghan Eq. (3) is the most EOS used [4]. This formalism describes very well the Bulk Modulus from an Energy x Volume curve. The point to point variations between the internal energy (U) and unit cell volume (V) build the curve. Posteriorly, the curve is fitted by Birch-Murnaghan Equation, which is a third degree polynomial.

$$U(V) = \frac{B_0 V}{B'_0}\left[\frac{\left(\frac{V_0}{V}\right)B'_0}{B'_0 - 1} + 1\right] + U_0 \tag{3}$$

where, U_0 is the total internal energy for the ground state, U(V) is the total internal energy for deformed unit cell, V_0 is the initial unit cell volume in the ground state, V is the deformed unit cell, and B_0 is the Bulk Modulus with B'_0 derivative.

All crystalline materials collapse from applied external pressure. Nevertheless, phase transitions are very essential to technological applications. In the area of the chemistry of materials, the phase transitions from applied external pressure are investigated to modify the electrical, optical, magnetic, and mechanical properties. In particular, Quantum simulations are largely applied to investigate modifications in crystalline structures under pressure. The phase transitions under pressure are simulated because of geophysics processes, piezoelectric technological devices, resistance materials, and magnetic phase transitions. The DFT increased in broad scale in the molecules and solids. DFT approaches for high pressure researches have a high level of publication and good agreement with experimental results. Recently, Quantum simulations are the source for energy (U) x unit cell volume (V) curve to be applied to EOS, for example, the Birch-Munaghan Equation. In codes based on Quantum formalism have implemented an automatic routine to calculate the *Bulk Modulus* and, posteriorly, the simulation for Gibb's x pressure curve showing phases transition analyses.

3. High pressure and density functional theory

A DFT approach to investigate the tetragonal-cubic phase transition of the $PbTiO_3$ was proposed to analysis the molecular connection between two crystalline structures [5]. The piezoelectric $PbTiO_3$ material is widely studied because of the high performance in sensors and other devices. It is knowledge that the tetragonal structure is a paraelectric phase with aggregation charges under an applied external pressure, the piezoelectric effect. The molecular structure is distorted through atomic displacement from the applied external pressure. In the $PbTiO_3$, there is the spontaneous polarization effect caused by the Ti atom displacement from the octahedral center inside $[TiO_6]$ cluster. Such molecular dynamic connects the Ti atomic displacement to the electric charging of the $PbTiO_3$ material from an anisotropic charge distribution. Simultaneously, the crystalline structure ways to cubic structure increasing the material resistance to applied external pressure, for each change in the applied external pressure there is a change in the electric signal of the $PbTiO_3$ material. Then, it was proposed DFT simulations to identify crystalline transition structures between tetragonal and cubic structures. The crystalline intermediary structures were featured showing a chemical path in consistence with the most recent concepts and schemes used in the solid state theory. The association between tetragonal-cubic phase transition and chemical path is very grateful presenting a high potential to clarify many molecular mechanisms associated to piezoelectric.

Another type of DFT simulation is the phase transition among more two crystalline structures [6]. One example, it was the phase transition investigated for seven SnO_2 crystalline structures. The rutile, $CaCl_2$, α-PbO_2, pyrite, ZrO_2, fluorite, and cotunnite symmetries were researched to demonstrate the changes in phase transition under pressure, structure, and optical properties. The calculated phase transitions showed a several phase transition points among symmetries with a high quantity of inter transitions, around seven interconversions structural, in a range of 30 GPa. A high malleability of this material was revealed and an increasing linear behavior for the band-gap in the 0 to 40 GPa range. From this expressive amount of calculations was possible to map many phase transitions and features for each crystalline structures.

The application of the external pressure on crystalline structures is intrinsically connected to density of the solid. From the increase of the applied pressure higher the density of solid; from Le Chatelier's Principle predicts that [7] "a decrease in volume accompanies first-order transitions from low pressure to high pressure"

Then, lower densities solids have lower Bulk Modulus because can be compressed to structures densest and ordered; whereas, the change in the Bulk Modulus for high densities solids is difficult because the crystalline structure does not support an increase on density or coordination number.

A DFT investigation using high pressure procedure researched a set of materials based on iron in the ABO_3 ilmenite structure [8]. Such materials are magnetics and the B cation (+4) was modified among 4A (Si, Ge, Sn) and 4B (Ti, Zr, Hf) group cations. The $FeBO_3$ materials showed an increase of density from Si to Hf; however, it was possible group it from similar densities, $FeSiO_3$ (4.72 g cm^{-3}) and $FeTiO_3$ (4.73 g cm^{-3}), $FeGeO_3$ (5.72 g cm^{-3}) and $FeZrO_3$ (5.27 g cm^{-3}); $FeSnO_3$ (6.41 g cm^{-3}) and $FeHfO_3$ (7.87 g cm^{-3}). The similarity is centered on the table periodic period showing the influence of the ionic radius on the density. The calculated Bulk Modulus showed that $FeZrO_3$, $FeGeO_3$ and $FeSiO_3$ are the most compressible materials with 12.38 GPa, 32.86 GPa, and 43.70 GPa, respectively. The $FeTiO_3$ is the most hard material for compression; while, the $FeSnO_3$ and $FeHfO_3$ materials are unstable because the calculated Bulk Modulus were − 731.76 GPa and − 142.94 GPa, respectively. Then, the lower materials in Bulk Modulus have possibility to change to other phases. For the $FeTiO_3$, the phase transition is hard because high Bulk Modulus; whereas, $FeSnO_3$ and $FeHfO_3$ materials have a high possibility to decompose to SnO_2, FeO, and HfO_2 simple oxides.

4. Conclusions

High pressures theoretical approaches are very used to simulate energetic stability and structural properties in several materials. The Bulk Modulus modification from chemical substitutions or crystalline structures is a versatile tool to predict some properties of a material. In oxide semiconductors simulations, the Bulk Modulus project the material's compressilibity for phase transition or piezoelectric applications. Electric, optical and magnetic properties can be associated to high pressures changes creating sensors devices.

Acknowledgements

I thanks to Araucária Foundation, CAPES, and Cnpq foment agencies; and also to the researchers from Chemistry Simulation Group.

Author details

Sérgio Ricardo de Lázaro
Department of Chemistry, Chemistry Simulation Group, State University of Ponta Grossa, Av. Gen. Carlos Cavalcanti, 4748, Zip code: 84030-900, Paraná, Brazil

*Address all correspondence to: srlazaro@uepg.br

References

[1] The Free Dictionary. [Internet]. Available from: https://encyclopedia2.thefreedictionary.com/Young%27s+moduli [Accessed: 2020-10-15]

[2] The Editors of Encyclopaedia Britannica. [Internet]. Available from: https://www.britannica.com/science/shear-modulus [Accessed: 2020-10-15].

[3] The Editors of Encyclopaedia Britannica. [Internet]. Available from: https://www.britannica.com/science/bulk-modulus [Accessed: 2020-10-15].

[4] MURNAGHAN, F. D. The Compressibility of Media under Extreme Pressures. Proceedings of the National Academy of Sciences of the U.S.A., v. 30, n. 9, p. 244-247, 1944

[5] RIBEIRO, R. A. P. and DE LÁZARO, S. R. theoretical investigations of the bulk modulus in the tetra-cubic transition of PbTiO3 material. Química Nova, v. 37, n. 7, p. 1165-1170, 2014

[6] GRACIA L, BELTRÁN A, ANDRÉS J. Characterization of the high-pressure structures and phase transitions in SnO_2. A density functional theory study. The Journal of Physical Chemistry. B. 2007;**111**:6479-6485

[7] WEST, A. R., Solid State Chemistry and its Applications. Chichester: Wiley, 1984. p. 445. ISBN: 0471903779

[8] RIBEIRO, R. A. P.; JR. CAMILO, A.; DE LAZARO, S. R. electronic structure and magnetism of new ilmenite compounds for spintronic devices: FeBO3 (B = Ti, Zr, Hf, Si, Ge, Sn). Journal of Magnetism and Magnetic Materials, v. 394, p. 463-469, 2015

Section 2

Chemical Reactivity

Chapter 2

Quantum Computational Chemistry: Modeling and Calculation of S-Block Metal Ion Complexes

Rakesh Kumar and Sangeeta Obrai

Abstract

The computational study of some s-block metal nitrophenolate complexes, $[Ca(THEEN)(PIC)]^+$ (**1**), $[Ca(THPEN)(H_2O)_2]^{2+}$ (**2**), $Ba(THPEN)(PIC)_2$ (**3**) $[Na(THPEN)]_2^{2+}$ (**4**), $[Sr(THPEN)(H_2O)_2]_2^{2+}$ (**5**) and $[Ba(THPEN)(H_2O)_2]_2^{2+}$ (**6**) (where THEEN (N,N,N′,N′-Tetrakis(2-hydroxyethyl)ethylenediamine) and THPEN (N,N,N′,N′-Tetrakis(2-hydroxypropyl)ethylenediamine) are tetrapodal ligands and PIC^- is 2,4,6-trinitrophenolate anion), is presented here using density functional theory (DFT) in its hybrid form B3LYP. The geometries of the title complexes are described by the quantum-chemical approach using input coordinates obtained from the previously synthesized and X-ray characterized diffraction data of [Ca(THEEN)(PIC)](PIC), $[Ca(THPEN)(H_2O)_2](PIC)_2$, $Ba(THPEN)(PIC)_2$, $[Na(THPEN)]_2(PIC)_2$, $[Sr(THPEN)(H_2O)_2]_2(DNP)_4$ and $[Ba(THPEN)(H_2O)_2]_2(DNP)_4$ (where DNP is 3,5-dinitrophenolate). Only the primary coordination sphere of complexes (**1–6**) is optimized in the gaseous phase. Calculations of the energy gaps of frontier orbitals (HOMO-LUMO), ^{13}C-NMR shifts and vibrational bands are carried out using B3LYP/6-31 g + (d,p)/LANL2DZ level of theory. The calculated geometric and spectral parameters reproduced the experimental data with a well agreement.

Keywords: DFT, s-block metal complexes, nitrophenolates, tetrapodal ligands

1. Introduction

The alkali and alkaline earth metal cations have inert gas electronic structures and are not expected to show any stereochemical requirements in their complex formation as do transition metal cations. They may be considered spherical even in the complex state. Their complexation is thus treated as recognition of spherical cations by organic ligands [1]. Depending upon the nature of the organic ligand and the anion, the metal ions can be separated as solvated ions, solvent separate, loose and tight ion pairs.

Alkali and alkaline earth metal ions form a large number of solid complexes with podands [2–6]. The podands are inherently flexible because the two ends of the molecule are not tied simultaneously. Polypodal ligands are acyclic multidentate ligands containing more than three arms. They form an unlimited family of structures which finally give rise to dendrimers. A predominant 1:1 complexation has also been observed in alkali and alkaline earth metal complexes of the tetra- and pentapodands. The stability constants of the terapodands are generally lower than those of the corresponding tripodands because of more severe steric hindrance to complexation [7]. Vögtle and coworkers have indicated that the ligands resembling tetrapodands are capable of forming 1:1 complexes with s-block metal ions [8, 9].

The s-block elements present a usual challenge in the molecular modeling, because the metal-ligand interactions in both cases are principally electrostatic. The types of alkali and alkaline earth metal complexes subjected to molecular modeling can be divided into five categories: crown ethers [10–16], cryptands [17, 18], spherands [19, 20], podands and other biologically important ligands, such as ionophores and cyclic antibiotics [21–24] . The present work has been undertaken with the aim to computationally characterize the structure and nature of complexes of s-block metal ions with the tetrapodands THEEN and THPEN. Recently the computational studies of these tetrapodal ligands with Cu(II), Ag(I) and La(III) have been reported. Recently, synthesis, crystal structure and biological properties of [Co(edtp)Cl]·NO_3·H_2O complex was also determined, where edtp is N,N,N′, N′-Tetrakis(2-hydroxypropyl) ethylenediamine in which Co^{2+} ion is coordinated by the N,N′,O,O′,O″-pentadentate edtp ligand and a chloride to generate a distorted $CoClN_2O_3$ octahedron [25–28].

2. Computational details

From the last 3 decades density functional theory has been the dominant method for the quantum mechanical simulation of periodic systems. In recent years it has also been adopted by quantum chemists and is now very widely used for the simulation of energy surfaces in molecules. The quantum-chemical calculations (DFT calculations) giving molecular geometries of minimum energies, molecular orbitals (HOMO-LUMO), ^{13}C-NMR and vibrational spectra were performed using the Gaussian 09 [29]. Molecular orbitals were visualized using "Gauss view". The method used was Becke's three-parameter hybrid-exchange functional, the nonlocal correlation provided by the Lee, Yang and Parr expression, and the Vosko, Wilk, and Nuair 1980 local correlation functional (III) (B3LYP) [30, 31]. The 6-31 g + (d,p) basis set was used for C, N and O. The LANL2DZ basis set [32] and pseudopotentials of Hay and Wadt were used for Ca, Sr, Ba and Na metal atoms [33, 34]. DFT calculations were performed in the gaseous phase and the input coordinates were obtained from and then compared with crystal structure data of already reported complexes: [Ca(THEEN)(PIC)](PIC),[Ca(THPEN)$(H_2O)_2$]$(PIC)_2$, Ba(THPEN)$(PIC)_2$, [Na (THPEN)]$_2$$(PIC)_2$, [Sr(THPEN)$(H_2O)_2$]$_2$$(DNP)_4$ and [Ba(THPEN)$(H_2O)_2$]$_2$$(DNP)_4$ (where DNP is 3,5-dinitrophenolate) [35]. The structural parameters were adjusted until an optimal agreement between calculated and experimental structure obtained throughout the entire range of available structures. HOMO-LUMO analyses and spectroscopic calculations were performed on the optimized geometries of the title complexes (**1–6**) using B3LYP/6-31 g + (d,p)/LANL2DZ level of theory.

3. Geometrical structures

3.1 Results and discussion

Complexes (**1–6**) were successfully modeled by using the input coordinates of crystal data. **Tables 1** and **2** represent comparison of calculated and experimental

	Complex 1 (M = Ca)			Complex 2 (M = Ca)			Complex 3 (M = Ba)		
	Theo.	Exp.	Dev.	Theo.	Exp.	Dev.	Theo.	Exp.	Dev.
M-O1	2.450	2.410	0.040	2.389	2.341	0.048	2.720	2.722	−0.002
M-O2	2.450	2.380	0.070	2.495	2.498	−0.003	2.807	2.812	−0.006
M-O3	2.410	2.480	0.070				2.753	2.753	0.000
M-O4	2.410	2.370	0.040				2.812	2.816	−0.004
M-O5							2.681	2.687	−0.006
M-O6							3.127	3.135	−0.008
M-O12	2.310	2.30	0.01				2.728	2.735	−0.007
M-O13	2.470	2.73	0.26						
M-O18							2.977	2.990	−0.013
M-O1W				2.440	2.442	−0.002			
M-O2W									
M-N1	2.824	2.591	0.233	2.601	2.603	−0.002	3.038	3.042	0.004
M-N2	2.738	2.658	0.08				3.026	3.032	−0.006

Table 1.
Comparison of experimental and calculated M-ligand bond lengths (Å) of complexes (1–3).

	Complex 4 (M = Na)			Complex 5 (M = Sr)			Complex 6 (M = Ba)		
	Theo.	Exp.	Dev.	Theo.	Exp.	Dev.	Theo.	Exp.	Dev.
M-O1	2.412	2.416	−0.004	2.617	2.628	−0.011	2.736	2.743	−0.007
M-O2	2.393	2.396	−0.003	2.611	2.618	−0.007	2.763	2.767	−0.004
M-O3	2.505	2.508	−0.003	2.506	2.516	−0.010	2.756	2.762	−0.006
M-O4	2.628	2.632	−0.004	2.618	2.626	−0.008	2.658	2.668	−0.010
M-O1W				2.701	2.711	−0.010	2.880	2.884	−0.004
M-O2W				2.699	2.705	−0.006	2.988	2.995	−0.007
M-O2WA				2.726	2.732	−0.006			
M-N1				2.835	2.842	−0.007	3.009	3.008	−0.001
M-N2				2.849	2.857	−0.008	3.010	3.010	0.000

Table 2.
Comparison of experimental and calculated M-ligand bond lengths (Å) of complexes (4–6).

M-Ligand bond lengths (Å) of complexes (**1**–**3**) and (**4**–**6**), respectively. The picrates and dinitrophenolates that are excluded from the primary coordination spheres of metal atoms in the crystallographic determinations, are not optimized in the computed structures. **Tables 3** and **4** represent comparison of calculated and experimental torsion angles of ligand in complexes (**1**–**3**) and (**4**–**6**) respectively.

3.1.1 $[Ca(THEEN)(PIC)]^+$ (1)

The coordination number of Ca(II) ion is eight with distorted cube geometry in the optimized geometry of cationic complex (**1**) (**Figure 1**). THEEN is interacting with Ca(II) ion through all the six potential donor atoms. The seventh and eighth coordination sites of Ca(II) are occupied by picrate anion through phenolic oxygen

Name of the	Complex (1)			Complex (2)			Complex (3)		
Atoms	**Theo.**	**Exp.**	**Dev.**	**Theo.**	**Exp.**	**Dev.**	**Theo.**	**Exp.**	**Dev.**
O1-C1-C2-N1	−55.1	−55.05	0.05	57.9	58.0	0.1	−32.3	−32.4	−0.1
C1-C2-N1-C5	165.1	165.1	0.0	82.8	82.7	0.1	161.9	162.0	0.1
C2-N1-C5-C6	−159.0	−159.0	0.0	−165.5	−165.5	0.0	−157.6	−157.8	−0.2
N1-C5-C6-N2	65.4	65.3	0.1	60.4	60.5	0.1	65.9	66.1	0.2
C5-C6-N2-C8				−159.1	−159.0	0.1	−159.3	−159.4	0.1
C6-N2-C8-C7				89.1	89.1	0.0	163.6	163.7	0.1
N2-C8-C7-O3				60.9	60.9	0.0	−34.1	−34.2	−0.1
O2-C3-C4-N1	36.5	36.7	0.2	55.5	55.6	0.1	46.8	46.9	−0.1
C3-C4-N1-C5	−121.6	−121.7	−0.1	−161.0	−161.0	0.0	−132.6	−132.7	−0.1
C4-N1-C5-C6	77.2	77.2	0.0	72.6	72.6	0.0	83.7	83.8	0.1
C5-C6-N2-C10				79.4	79.4	0.0	79.0	79.1	0.1
C6-N2-C10-C9				−151.8	151.8	0.0	−134.9	−134.9	0.0
N2-C10-C9-O4				61.3	61.4	0.1	52.1	52.2	0.1

Table 3.
Comparison of calculated and experimental torsion angles (°) of ligand in the complexes (1–3).

Name of the	Complex (4)			Complex (5)			Complex (6)		
Atoms	**Theo.**	**Exp.**	**Dev.**	**Theo.**	**Exp.**	**Dev.**	**Theo.**	**Exp.**	**Dev.**
Complex	−55.6	−55.7	−0.1	32.3	32.3	0.0	−43.4	−43.5	0.1
O1-C1-C2-N1	152.7	152.6	0.1	−143.9	−143.9	0.0	162.3	162.4	0.1
C1-C2-N1-C5	−80.2	−80.2	0.0	86.1	86.1	0.0	−92.7	−92.8	0.1
C2-N1-C5-C6	−64.0	−63.9	−0.1	43.6	43.6	0.0	−50.1	−50.1	0.0
N1-C5-C6-N2	165.5	165.4	0.1	−146.4	−146.4	0.0	156.5	156.7	0.2
C5-C6-N2-C8	−85.7	−85.7	0.0	156	155.9	0.1	−89.6	−89.9	0.3
C6-N2-C8-C7	−60.3	−60.4	−0.1	−44.7	−44.7	0.0	−30.4	−30.4	0.0
N2-C8-C7-O3	−59.7	−59.8	−0.1	−17.4	−17.4	0.0	47.8	48.1	0.3
O2-C3-C4-N1	−91.0	−91.1	−0.1	112.7	112.7	0.0	−135.1	−135.2	0.1
C3-C4-N1-C5	156.0	156.0	0.0	−159.1	−159.2	0.1	151.9	151.9	0.0
C4-N1-C5-C6	−72.4	−72.5	−0.1	97.4	97.4	0.0	−92.1	−92.4	0.3
C5-C6-N2-C10	156.6	156.6	0.0	−151.8	−151.8	0.0	143.4	143.9	0.5
C6-N2-C10-C9	−59.2	−59.3	−0.1	41.6	41.7	0.1	5.1	4.9	0.3
N2-C10-C9-O4	−55.6	−55.7	−0.1	32.3	32.3	0.0	−43.4	−43.5	0.1

Table 4.
Comparison of experimental and calculated and torsion angles (°) of ligand in the complexes (4–6).

and one of the o-nitro oxygen. The observed and calculated positions of calcium and donor atoms are in agreement. A comparison of bond lengths and bond angles provided a maximum tolerance of 0.27 Å and 16.33°Å, respectively (**Table 1** and **Table S1**; **Figure S1a,b**) of computed title complex (**1**) and crystallographically determined complex [Ca(THEEN)(PIC)](PIC). The torsion angles of the ligand THEEN in theoretically determined and experimental complex are also in well

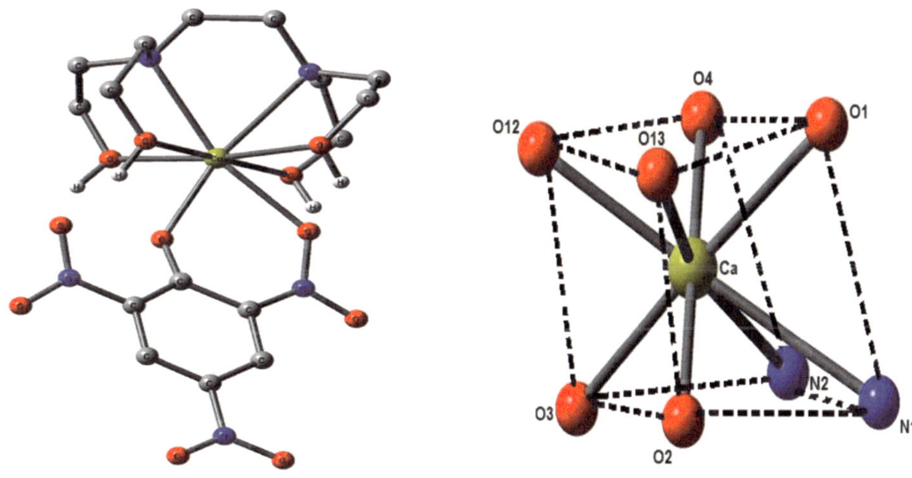

(a) (b)

Figure 1.
(a) Optimized geometric structure of [Ca(THEEN)(PIC)]$^{+}$ (1) (b) distorted cube geometry.

consistency with each other (**Table 3**). The HOMO-LUMO analysis has indicated that there is maximum distribution of HOMO over the carbon atoms and to a lesser extent on the oxygen of the coordinated picrate. LUMO is mainly distributed or concentrated on all the atoms of coordinated picrate except p-nitro group. Neither HOMO nor LUMO distribution is found either on calcium ion or on any ligand atom. The HOMO-LUMO gap (ΔE) is found to be 0.852 eV (**Figure S1c**).

3.1.2 [Ca(THPEN)(H_2O)$_2$]$^{2+}$ (2)

Ca(II) is eight coordinated in monomeric cationic complex with a distorted square-antiprismatic geometry in complex (2) as is observed in crystal structure of [Ca(THPEN)(H_2O)$_2$] (PIC)$_2$. H_2O (**Figure 2**). THPEN is acting as hexadentate ligand towards the metal ion. Two remaining sites around the metal ion are occupied by two water molecules. The observed and calculated positions of calcium and donor atoms are in agreement. All strain energy minimized structures reproduced the observed X-ray structures with almost no deviation in M-L bond length and torsion angle of the ligand (**Table 3**). The maximum tolerance of L-M-L bond angle is 15.8° (**Tables 1** and **Table S1, Figure S2a,b**). Complex 2 is displaying the main distribution of HOMO as well as LUMO only on water molecules. There is slight distribution of HOMO and LUMO on hydroxyl oxygen of THPEN ligand also with HOMO-LUMO gap (HLG) of 0.419 eV (**Figure S2c**).

3.1.3 Ba(THPEN)(PIC)$_2$ (3)

Ba(II) is ten-coordinate in its monomeric complex (3) (**Figure 3**). The ligand THPEN coordinates through both its nitrogen atoms and all the four hydroxyl oxygen. The two picrate ions are also directly coordinated to Ba^{2+} ion through the phenolic oxygen and one oxygen of the o-nitro group. Since both the picrates are directly interacting with the cation, the complex is termed as tight ion-paired complex as is found in the crystallographically determined complex. The observed and calculated positions of barium and donor atoms are in agreement. There is negligible deviation found in bond length, bond angle (0.11°) and torsion angle of ligand THPEN for calculated title complex (3) and experimental Ba(THPEN)(PIC)$_2$

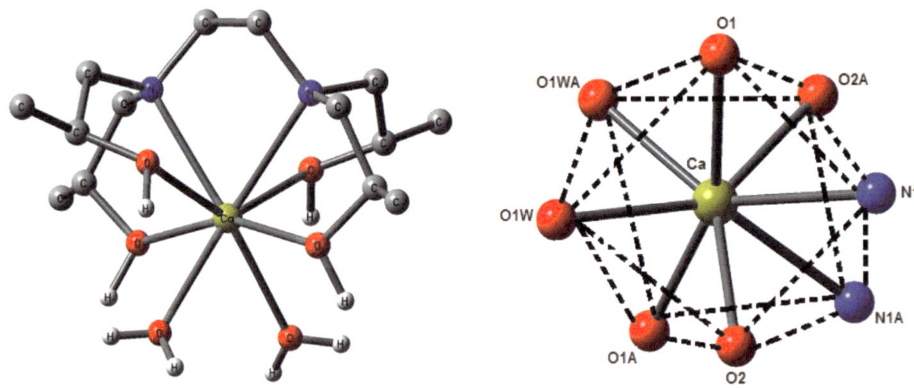

(a) (b)

Figure 2.
(a) Optimized geometric structure of $[Ca(THPEN)(H_2O)_2]^{2+}$ *(2) (b) distorted square-antiprismatic geometry.*

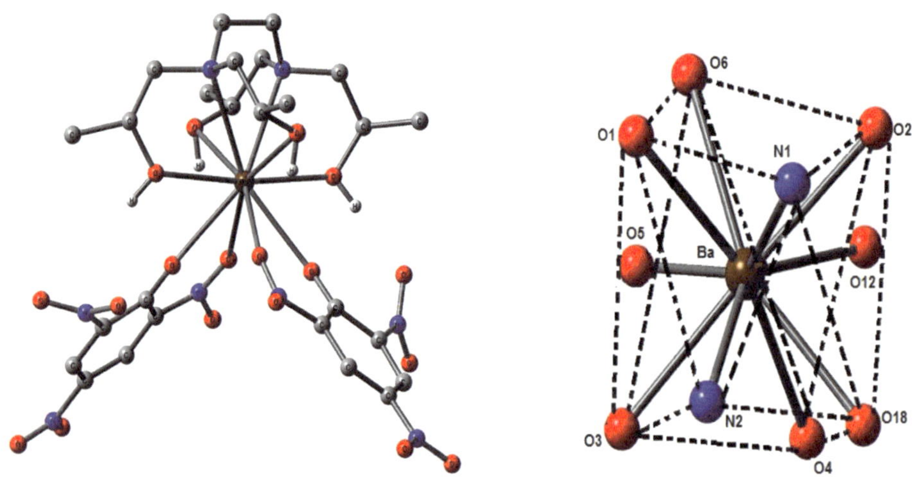

(a) (b)

Figure 3.
(a) Optimized geometric structure of $Ba(THPEN)(PIC)_2$ *(3) (b) bicapped square-antiprismatic geometry.*

(**Tables 1** and **3** and S1, **Figure S3a,b**). An analysis of HOMO-LUMO has illustrated that HOMO is mainly distributed on the ligand THPEN with a small distribution on coordinated picrates. In contrast to this LUMO is mainly distributed on coordinated picrates with no distribution over the ligand or metal atom. The complex is showing HLG of 0.118 eV (**Figure S3c**).

3.1.4 $[Na(THPEN)]_2^{2+}$ *(4)*

Figure 4a shows the optimized structure of the cationic complex of sodium (**4**). This structure is centrosymmetric dimer. The coordination number around each Na^+ ion is seven. Each THPEN ligand coordinates to Na(I) ion in a heptadenate fashion (**Figure 4**). In other words, one hydroxyl group of the THPEN ligand acts as a bridge between two Na^+ ions. The geometry of the complex is distorted monocapped octahedron. The Na ... Na non bridging distance is 3.429 Å. Almost no

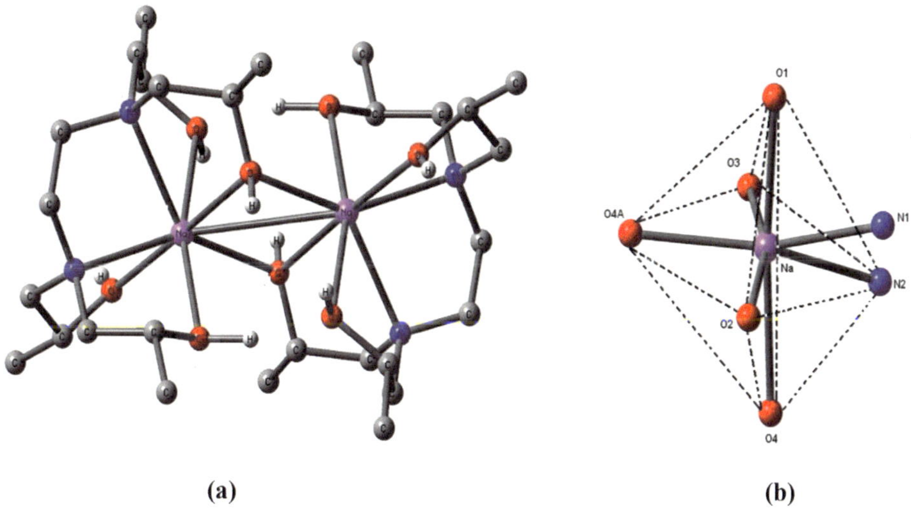

Figure 4.
(a) Optimized geometric structure of $[Na(THPEN)]_2^{2+}$ (4) (b) distorted monocapped octahedron geometry.

deviation has been observed in bond length (M-L), torsion angle of ligand THPEN and bond angles (L-M-L) of computed title complex (**4**) and crystallographically determined complex $[Na(THPEN)]_2(PIC)_2$ (**Tables 2** and **4** and **Table S1, Figures S4a,b**). The HOMO-LUMO study has revealed that in the title complex (**4**) HOMO is mainly concentrated on bridged hydroxyl oxygens and sodium metal but to a smaller extent on the other coordinated hydroxyl oxygens and amine nitrogens. LUMO is mainly concentrated on bridged hydroxyl oxygens and to a smaller extent on other coordinated oxygen atoms. The HOMO-LUMO gap is 0.261 eV (**Figure S4c**).

3.1.5 $[Sr(THPEN)(H_2O)_2]_2^{2+}$ (5)

Sr(II) is nine-coordinated in its dimeric complex (5) (**Figure 5**). This cationic complex is having tricapped trigonal-prismatic geometry as is found in crystallographically determined complex $[Sr(THPEN)(H_2O)_2]_2(DNP)_4$. Each Sr^{2+} ion in the complex is coordinated by six potential donor sites of the ligand THPEN and four water molecules. Out of four water molecules, two are bridging and the third one is non-bridging. Sr ... Sr non bridging distance is 4.346 Å indicating the existence of van der Waals contact. All strain energy minimized structures reproduced the observed X-ray structures to a maximum tolerance of 9.07° bond angle (**Table 2** and **Table S1, Figure S5a,b**). Almost no deviation of M-L bond length and torsion angle of ligand THPEN has been observed for the title complex (5) and crystallographically determined complex (**Tables 2** and **4**). The HOMO-LUMO analysis has shown that the complex (5) is having maximum distribution of HOMO and LUMO on bridged coordinated water molecules and there is HLG of 0.0225 eV (**Figure S5c**).

3.1.6 $[Ba(THPEN)(H_2O)_2]_2^{2+}$ (6)

Ba(II) is ten-coordinate in the cationic title complex (**6**) as is found in the crystallographic determined complex $[Ba(THPEN)(H_2O)_2](DNP)_4$. The geometry around Ba(II) is bicapped cubic. Each Ba(II) in dimer is interacting with THPEN through all its six potential donor sites and four water molecules. The latter are bridging in nature. The existence of van der Waals contact between non-bridging Ba ... Ba is indicated by their larger distance (4.196 Å). The observed and calculated positions of the metal and donor atoms are in agreement as almost negligible deviation of M-L bond length and torsion angle of THPEN. A deviation of 0.14° of bond angle L-M-L has been observed for complex (**6**) [Ba(THPEN)$(H_2O)_2$](DNP)$_4$ (**Tables 2** and **4** and **Table S1**; **Figures S6a,b** and **S7**). The region of distribution of HOMO and LUMO is only over two of the bridged water molecules in complex (**6**) with a very small HOMO-LUMO gap (ΔE = 0.0375 eV) indicating the soft nature of complex (**Figure 6**).

3.2 HOMO-LUMO analyses

Smaller is the HUMO-LUMO gap (HLG) softer is the complex [36, 37]. The frontier orbitals HOMO and LUMO are very important parameters for chemical reaction and take part in chemical stability [38–40]. It is predicted from the HOMO-LUMO gaps that the title complexes are soft as is obvious from their smaller HUMO-LUMO energy gaps (HLG) relative to the similar reported complexes of

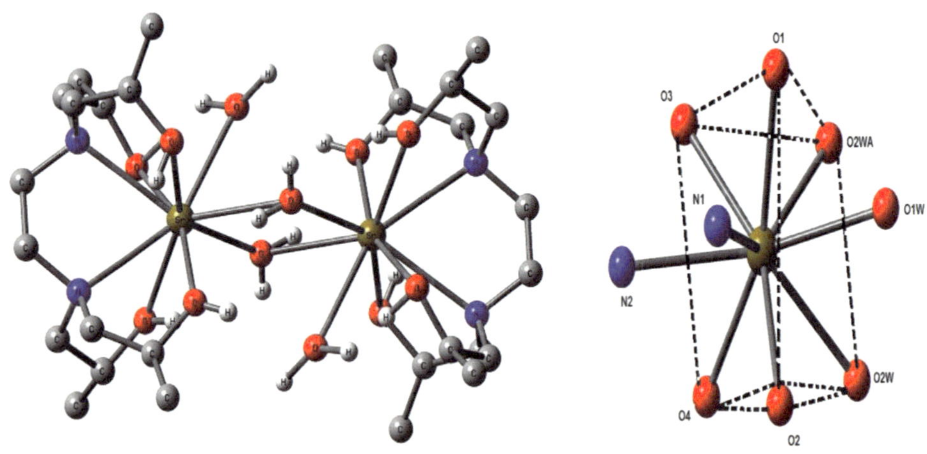

(a) (b)

Figure 5.
(a) Optimized geometric structure of [Sr(THPEN)$(H_2O)_2]_2$(DNP)$_4$ (b) Trigonal- prismatic geometry.

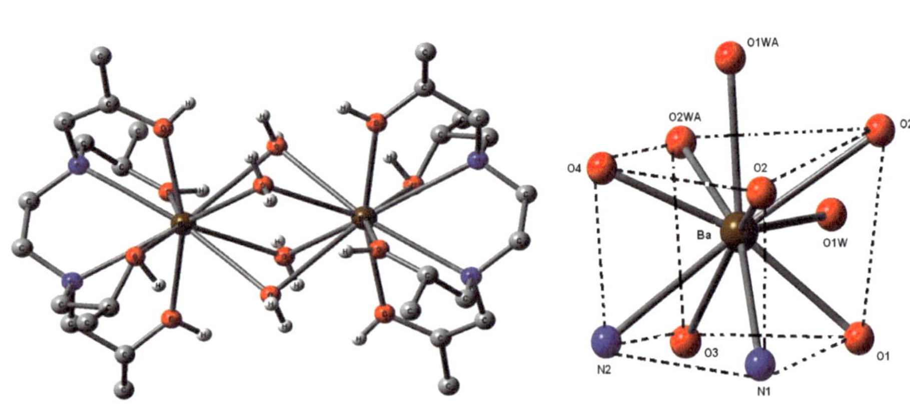

(a) (b)

Figure 6.
*(a) Optimized geometric structure of [Ba(THPEN)$(H_2O)_2]_2^{2+}$ (**6**) (b) Bicapped cubic geometry.*

copper, silver and lanthanoid [25–27] (**Table 5**). It has been observed in the present computational study that the dinitrophenolate complexes are softer than trinitrophenolate and among the latter, [Ca(THEEN)(PIC)]$^+$ (**1**) is displaying least softness. It is pertinent to mention here that the complex (**1**) is having THEEN ligand whereas THPEN is ligand in rest of the complexes (**2–6**).

3.3 Spectral data

Nuclear magnetic resonance spectra (NMR) and infrared (IR) spectroscopy can be useful for studying the coordination of various ligating sites. The ^{13}C-NMR spectra were predicted for complexes (**1–6**) using DFT/B3LYP/6-31G** method and the spectral data was compared with experimental data reported in literature [35]. The computed NMR spectral data is fairly in agreement with experimental data (**Tables 6** and 7). Small deviations are due to the fact that H-bonding interactions or any type of lattice interactions are not modeled in theoretically computed structures. The terminal methyl groups in theoretically predicted complexes (**2–5**) are

	HLG (eV)	Reported complexes	HLG (eV)	Reported complexes	HLG (eV)
[Ca(THEEN)(PIC)]$^+$ (**1**)	0.852	[Cu(THEEN)(H_2O)] $(PIC)_2$	3.537	[La(THEEN)(PIC) $(H_2O)_2$] $(PIC)_2.2H_2O$	3.428
[Ca(THPEN) $(H_2O)_2$]$^{2+}$ (**2**)	0.419	[Cu(THPEN)] $(PIC)_2.C_3H_8O$	3.467	[La($TEAH_3$)$(H_2O)_2$] $(PIC)_3$	3.673
Ba(THPEN)$(PIC)_2$ (**3**)	0.118	[Cu($TEAH_3$)(PIC)] (PIC).H_2O	3.619		
$[Na(THPEN)]_2^{2+}$ (**4**)	0.261	[Ag (THEEN)]$_2$ $(PIC)_2$	2.530		
[Sr(THPEN) $(H_2O)_2]_2^{2+}$ (**5**)	0.0225	[Ag (THPEN)]$_2$ $(PIC)_2$	2.640		
[Ba(THPEN) $(H_2O)_2]_2^{2+}$ (**6**)	0.0375	[Ag$(TEAH_3)_2$] (PIC)	1.061		

Table 5.
Comparison of HOMO-LUMO energy gaps of complexes (1–6) with earlier reported complexes [25–27].

Assignments (δ)	[Ca(THEEN)(PIC)]$^+$ (1)		Ba(THPEN)$(PIC)_2$ (3)	
	Theo.	Exp.	Theo.	Exp.
—CH_3	*	*	4.22	19.61
—CH_3			4.49	19.89
—NCH_2	52.31	55.03	51.73	55.34
—NCH_2			43.35	55.20
—OCH_2, —OCH	59.03	57.47	50.75	63.00
—ArCH	113.63	124.33	117.85	124.36
p-ArCN	134.75	123.60	122.06	123.68
o-ArCN	140.68	140.78	140.98	140.78
—ArCO	161.57	160.20	150.21	160.20

* *Group absent.*

Table 6.
Comparison of calculated and experimental ^{13}C-NMR spectral data for complexes (1 and 3).

Assignment (δ)	$[Ca(THPEN)(H_2O)_2]^{2+}$ (2)		$[Na(THPEN)]_2^{2+}$ (4)		$[Sr(THPEN)(H_2O)_2]_2^{2+}$ (5)		$[Ba(THPEN)(H_2O)_2]_2^{2+}$ (6)	
	Theo.	Exp.	Theo.	Exp.	Theo.	Exp.	Theo.	Exp.
$—CH_3$	11.35	18.79	4.01	20.30	3.77	18.31	27.02	18.35
$—CH_3$	11.38	19.02	5.63	20.36	3.85	18.59	27.69	18.63
$—NCH_2$	39.60	50.99	43.05	50.42	43.60	48.31	61.69	59.80
$—NCH_2$	39.61	51.76	44.05	52.90	57.81	59.68	61.69	60.59
—OCH	*	*	50.73	55.73	41.85	60.62	60.37	60.66
—OCH	69.24	61.74	57.90	55.95	51.91	61.08	60.38	62.53

* *Group absent.*

Table 7.
Comparison of calculated and experimental ^{13}C-NMR spectral data for complexes (2, 4–6).

Assignments (cm^{-1})	$[Ca(THEEN)(PIC)]^+$ (1)		$[Ca(THPEN)(H_2O)_2]^{2+}$ (2)		$Ba(THPEN)(PIC)_2$ (3)	
	Theo.	Exp.	Theo.	Exp.	Theo.	Exp.
ν (NO_2)	1363.56	1360 m	1320.81	1360 vs	1384.20	1370
δ (=CH)	696.80	700 m	784.16	790 m	800	800

Table 8.
Comparison of calculated and experimental IR spectral data for complexes (1–3).

Assignments (cm^{-1})	$[Na(THPEN)]_2^{2+}$ (4)		$[Sr(THPEN)(H_2O)_2]_2^{2+}$ (5)		$[Ba(THPEN)(H_2O)_2]_2^{2+}$ (6)	
	Theo.	Exp.	Theo.	Exp.	Theo.	Exp.
ν (NO_2)	1366	1371.21	1330	1329.24	1340	1329.43
δ (=CH)	790	784.54	760	762.52	790	787.17

Table 9.
Comparison of calculated and experimental IR spectral data for complexes (4–6).

displaying quiet high upfield shifts relative to the experimentally obtained due to more free movements in the gaseous phase than in solution or solid phase.

The computed IR spectral peaks that appear in the range of 1370–600 cm^{-1} are fairly in agreement with the experimental data (**Tables 8** and **9**). The absorption peaks due to the presence of hydroxyl groups were observed only for complexes (**1**) and (**6**) in the computed IR spectra. It is pertinent to mention here that both of these complexes possess the picrate anion in their coordination sphere. The less extent of H-bonding is reported in the crystallographic description of these complexes [35]. The theoretical absorption band appears at 3223 and 3235 cm^{-1} for both complexes (**1**) and (**6**) whereas the experimental band is reported at 3300 cm^{-1} for both of them [35].

4. Conclusions

The coordination number of the title s-block complexes is varying from 7 to 10 in the present work. As the size of the metal increases, coordination number of central metal ion also increases. Out of the six complexes presented, three are monomeric (**1–3**) and three are dimeric (**4–6**). The longer distances between two M ... M distances in the dinuclear complexes indicate the existence of van der Waals contacts between the two s-block metal ions. All the title complexes are cationic except Ba(THPEN)$(PIC)_2$ (**3**). The latter is tight ion-paired complex. All strain energy minimized structures obtained using quantum-chemical approach reproduced the observed X-ray structures with geometric parameters in well agreement. HOMO-LUMO studies suggest the softness of the title s-block complexes relative to the similar already reported copper, silver and lanthanoid complexes. The theoretical spectral data (^{13}C-NMR and IR) computed using DFT and experimental data is fairly in agreement with each other. The accuracy of the results predicts that the DFT studies performed using B3LYP/6-31 g + (d,p)/LANL2DZ level of theory is the appropriate quantum-chemical method for reproducing the experimental results for the title s-block complexes. This quantum-chemical approach has potential for molecular modeling of other s-block complexes and exploring their chemistry.

Small deviations in geometric as well as spectral parameters may be attributed to the lack of H-bonding and packing interactions within lattice which were not modeled during the computational study of the entitled s-block complexes. Moreover, the quantum-chemical approach of DFT studies has been carried out in the gaseous phase whereas the already reported experimental crystal and IR spectral data is in the solid phase while ^{13}C-NMR spectral data is in the solution phase.

Appendix

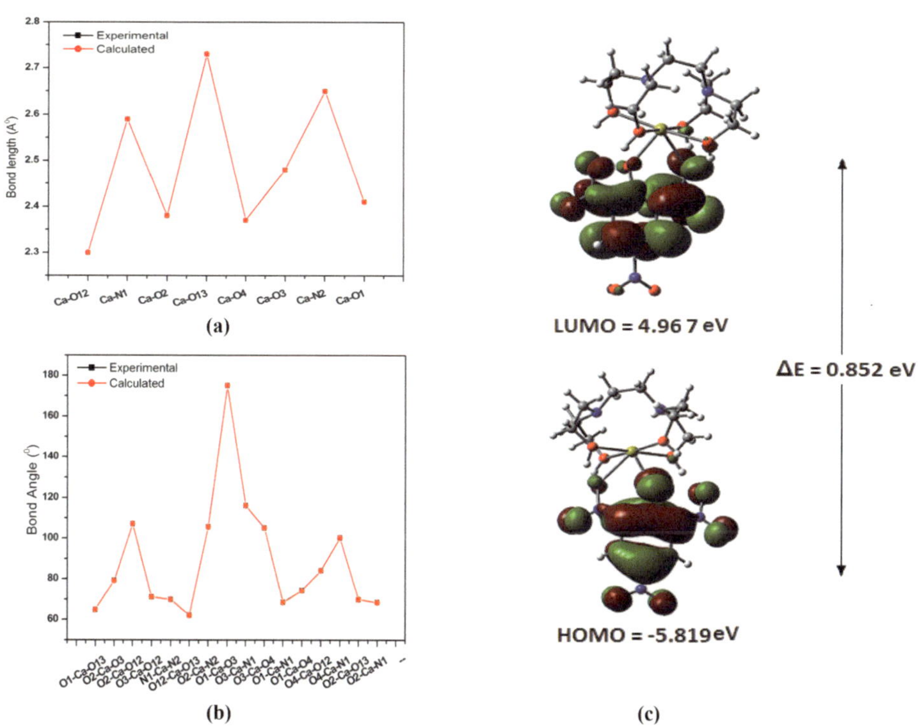

Figure S1.
Plot showing the deviations of theoretical and experimental (a) bond lengths (Å) and (b) bond angles (°) for the complex (1) and (c) (HOMO-LUMO) of the complex (1) with energy gap.

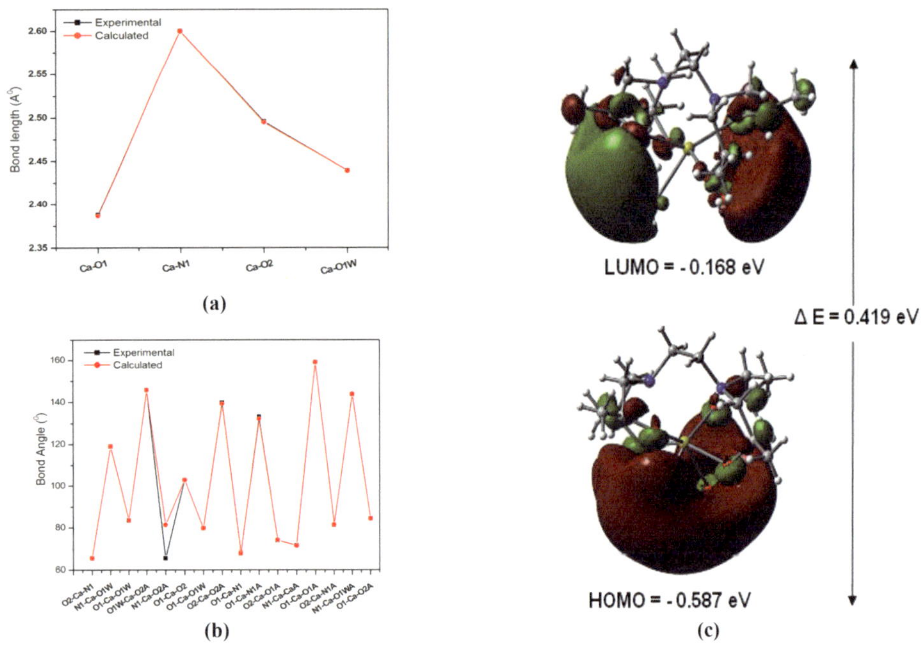

Figure S2.
Plot showing the deviations of theoretical and experimental (a) bond lengths (Å) and (b) bond angles (°) for the complex (2) (c) (HOMO-LUMO) of the complex (2) with energy gap.

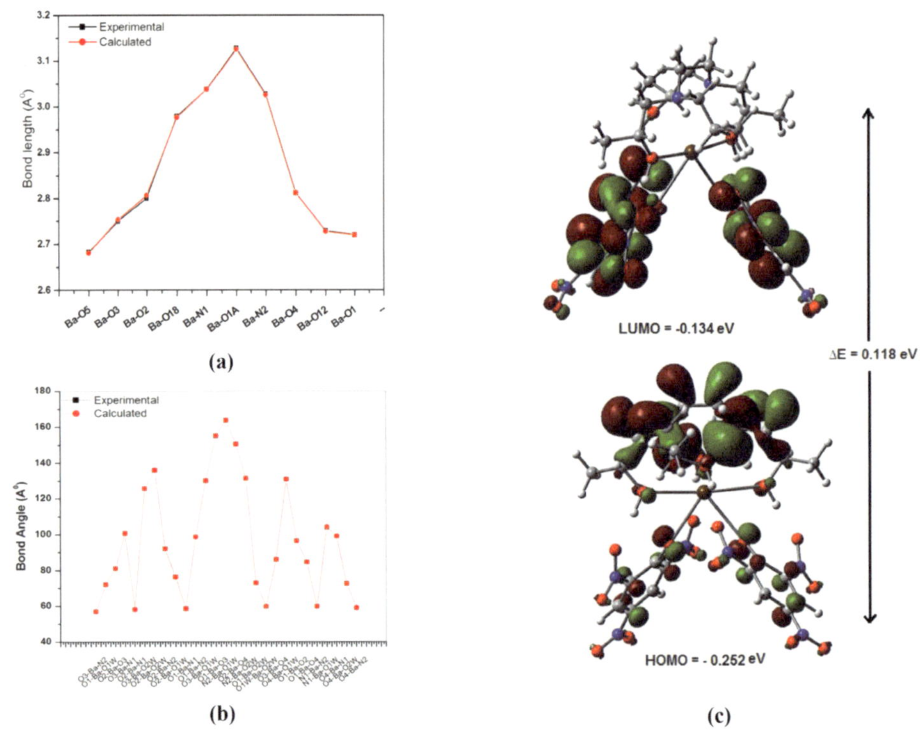

Figure S3.
Plot showing the deviations of theoretical and experimental (a) bond lengths (Å) and (b) bond angles (°) for the complex (3) (c) (HOMO-LUMO) of the complex (3) with energy gap.

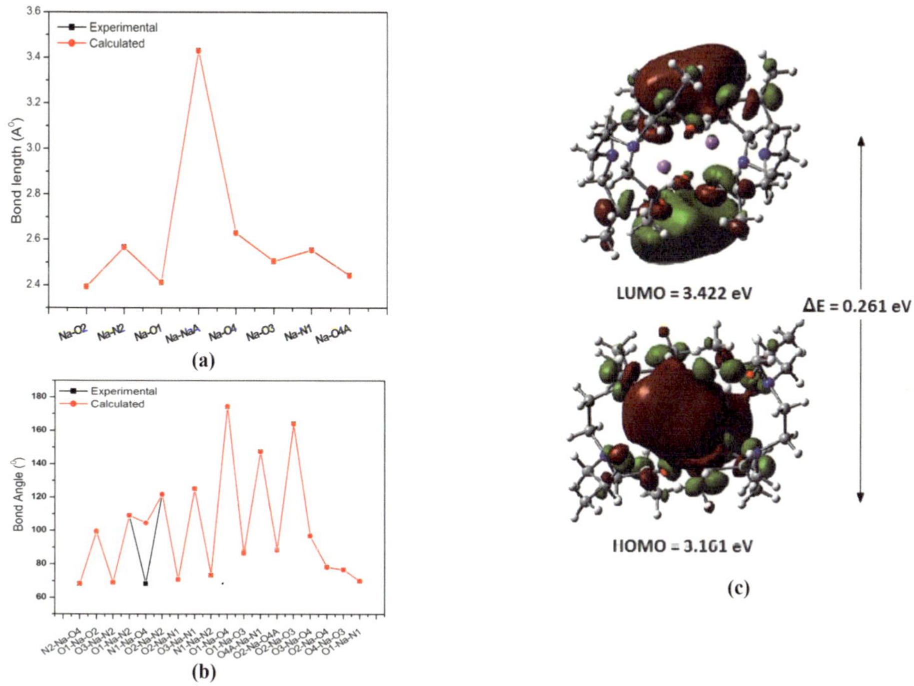

Figure S4.
*Plot showing the deviations of theoretical and experimental (a) bond lengths (Å) and (b) bond angles (°) for the complex (**4**) (c) (HOMO-LUMO) of the complex (**4**) with energy gap.*

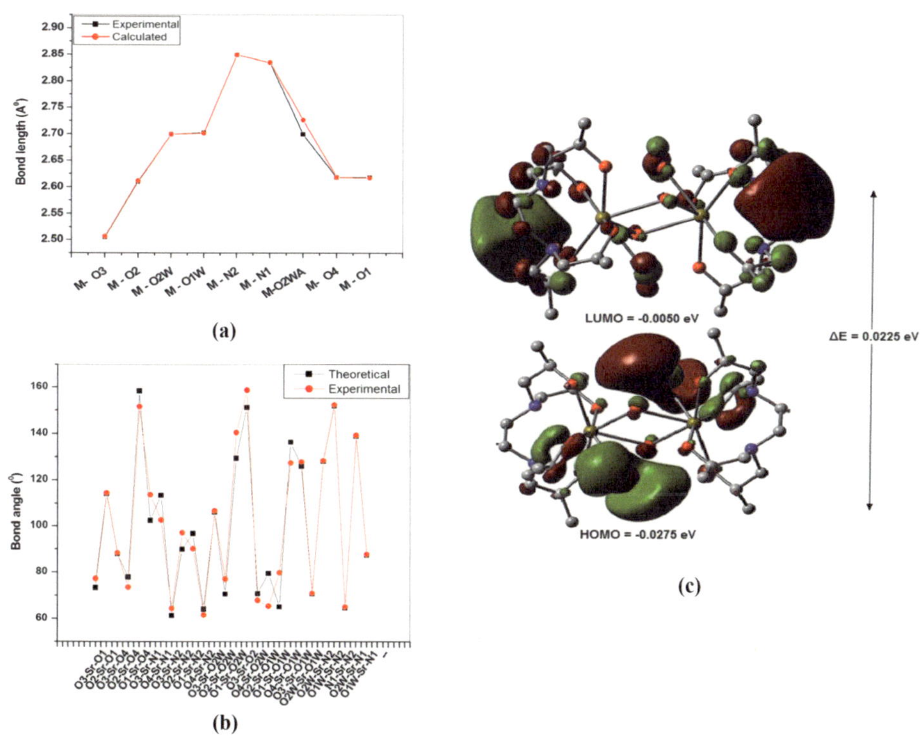

Figure S5.
Plot showing the deviations of theoretical and experimental (a) bond lengths (Å) and (b) bond angles (°) for the complex (5) (c) (HOMO-LUMO) of the complex (5) with energy gap.

v

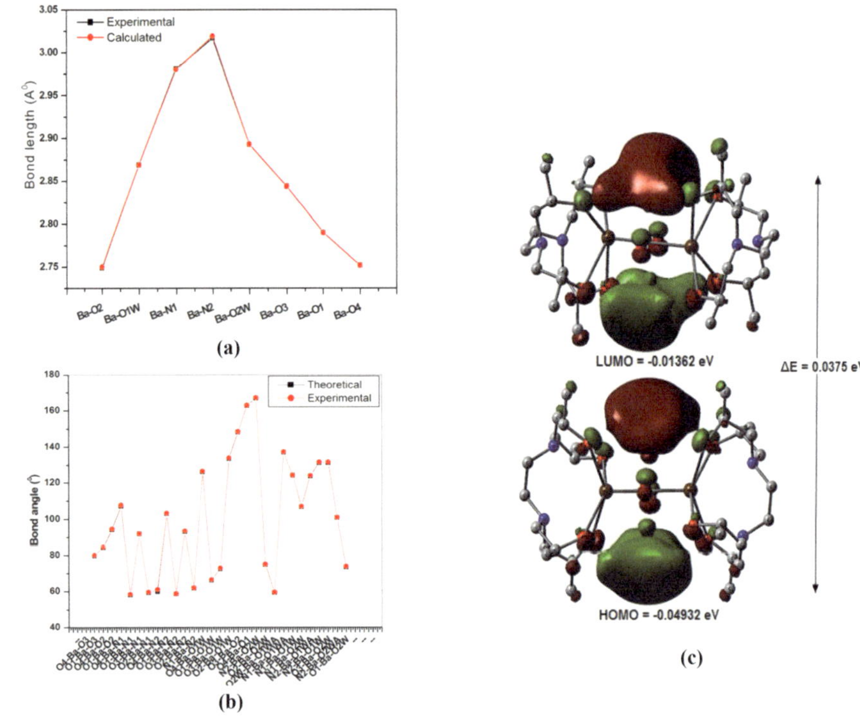

Figure S6.
*Plot showing the deviations of theoretical and experimental (a) bond lengths (Å) and (b) bond angles (°) for the complex (**6**) (c) (HOMO-LUMO) of the complex (**6**) with energy gap.*

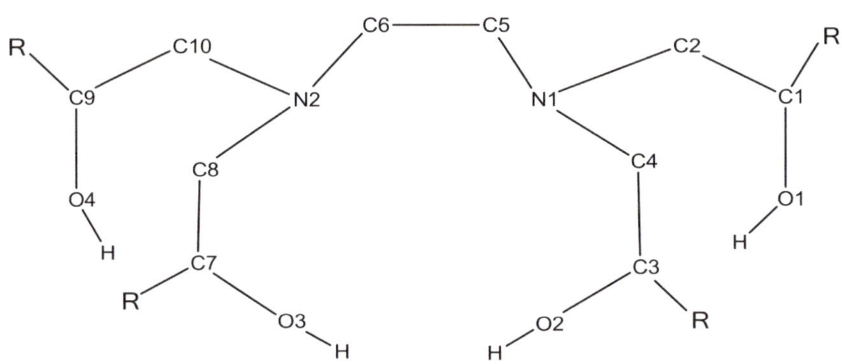

R = H, THEEN ; R = CH_3, THPEN

Figure S7.
R = H, THEEN; R = CH_3, THPEN.

Bond distances (Å)	Theoretical	Experimental	Dev.	Bond angles (°)	Theoretical	Experimental	Dev.
Complex (**1**)							
Ca-N1	2.600	2.600	0.000	N1-Ca-N1A	71.54	71.60	0.06
Ca-O1	2.387	2.388	0.001	N1-Ca-O2A	81.41	65.52	15.89
Ca-O2	2.495	2.496	0.001	N1-Ca-O1W	119.08	119.09	0.01

Bond distances (Å)	Theoretical	Experimental	Dev.	Bond angles (°)	Theoretical	Experimental	Dev.
Ca-O1W	2.439	2.439	0.000	N1-Ca-O1WA	143.89	143.89	0.00
				O1-Ca-O2	102.93	102.95	0.02
				O1-Ca-N1	67.80	67.79	0.01
				O1-Ca-O1W	79.82	79.85	0.03
				O1-Ca-N1A	132.22	132.22	0.00
				O1-Ca-O2A	84.37	84.36	0.01
				O1-Ca-O1A	159.11	159.10	0.01
				O1-Ca-O1WA	83.52	83.54	0.02
				O2-Ca-N1	65.50	65.52	0.02
				O2-Ca-O1W	74.11	74.10	0.01
				O2-Ca-N1A	81.42	81.40	0.02
				O2-Ca-O2A	139.33	139.89	0.56
				O2A-Ca-O1W	145.88	145.90	0.02
Complex (2)							
Ca-O1	2.45	2.41	0.04	O1-Ca-O2	94.43	102.08	7.65
Ca-O2	2.45	2.38	0.07	O1-Ca-O3	172.00	175.21	3.21
Ca-O3	2.41	2.48	0.07	O1-Ca-O4	85.56	74.35	11.21
Ca-O4	2.41	2.37	0.04	O1-Ca-O12	107.80	104.18	3.62
Ca-O12	2.31	2.30	0.01	O1-Ca-O13	61.91	64.85	2.94
Ca-O13	2.47	2.73	0.26	O1-Ca-N1	65.93	68.50	2.57
Ca-N1	2.74	2.59	0.15	O1-Ca-N2	104.33	115.49	11.16
Ca-N2	2.82	2.65	0.17	O2-Ca-O3	90.99	79.27	11.72
				O2-Ca-O4	180.00	168.65	11.35
				O2-Ca-O12	108.06	107.20	0.86
				O2-Ca-O13	61.38	69.92	8.54
				O2-Ca-N1	64.20	68.43	4.23
				O2-Ca-N2	114.39	105.72	8.67
				O3-Ca-O4	89.01	105.24	16.23
				O3-Ca-O12	75.95	71.06	4.89
				O3-Ca-O13	126.03	110.69	15.34
				O3-Ca-N1	111.54	116.18	4.64
				O3-Ca-N2	67.97	68.26	0.29
				O4-Ca-O12	71.93	84.20	12.27
				O4-Ca-O13	118.62	116.91	1.71
				O4-Ca-N1	115.80	100.38	15.42
				O4-Ca-N2	65.61	67.36	1.75
				O12-Ca-O13	71.32	62.19	9.13
				O12-Ca-N1	168.56	169.55	0.99
				O12-Ca-N2	123.47	120.54	2.93
				O13-Ca-N1	97.26	107.46	10.2

Bond distances (Å)	Theoretical	Experimental	Dev.	Bond angles (°)	Theoretical	Experimental	Dev.
				O13-Ca-N2	163.73	175.60	11.87
				N1-Ca-N2	67.95	69.92	1.97
Complex (**3**)							
Ba-O1	2.720	2.720	0.000	O1-Ba-O2	86.95	87.00	0.05
Ba-O2	2.807	2.801	0.006	O1-Ba-O3	172.87	173.00	0.13
Ba-O3	2.753	2.750	0.003	O1-Ba-O4	88.55	88.50	0.05
Ba-O4	2.812	2.812	0.000	O1-Ba-O6	72.90	72.90	0.00
Ba-O5	2.681	2.683	0.002	O1-Ba-O12	61.79	61.80	0.01
Ba-O6	3.127	3.128	0.001	O1-Ba-O18	109.63	109.70	0.07
Ba-O12	2.728	2.729	0.001	O1-Ba-N1	58.12	58.10	0.02
Ba-O18	2.977	2.979	0.002	O1-Ba-N2	114.85	114.80	0.05
Ba-N1	3.038	3.038	0.000	O2-Ba-O4	129.37	129.40	0.03
Ba-N2	3.026	3.028	0.002	O2-Ba-O6	64.79	64.80	0.01
				O2-Ba-O18	161.62	160.60	0.02
				O2-Ba-N2	79.36	79.40	0.03
				O2-Ba-N1	57.03	57.12	0.05
				O3-Ba-O2	92.08	92.10	0.02
				O3-Ba-O4	86.61	86.70	0.09
				O3-Ba-O6	113.02	113.10	0.08
				O3-Ba-O18	72.70	72.80	0.10
				O3-Ba-N1	115.63	115.60	0.03
				O3-Ba-N2	58.06	58.20	0.14
				O4-Ba-O6	156.82	156.80	0.02
				O4-Ba-O18	63.13	63.20	0.07
				O4-Ba-N1	78.20	78.20	0.00
				O4-Ba-N2	57.43	57.40	0.03
				O5-Ba-O1	123.12	123.20	0.08
				O5-Ba-O2	85.94	85.90	0.04
				O5-Ba-O3	63.80	63.90	0.10
				O5-Ba-O4	135.92	135.90	0.02
				O5-Ba-O6	53.26	53.30	0.03
				O5-Ba-O12	84.0	84.00	0.00
				O5-Ba-O18	76.67	76.90	0.13
				O5-Ba-N1	142.90	142.90	0.00
				O5-Ba-N2	118.97	118.90	0.07
				O12-Ba-O2	133.15	133.20	0.05
				O12-Ba-O3	123.02	123.00	0.02
				O12-Ba-O4	86.45	86.40	0.05
				O12-Ba-O6	72.69	72.70	0.01
				O12-Ba-O18	54.05	54.00	0.05

Bond distances (Å)	Theoretical	Experimental	Dev.	Bond angles (°)	Theoretical	Experimental	Dev.
				O12-Ba-N1	117.95	117.96	0.01
				O12-Ba-N2	143.78	143.80	0.02
				O18-Ba-O6	109.54	109.60	0.06
				O18-Ba-N1	140.27	140.40	0.13
				O18-Ba-N2	101.49	101.50	0.01
				N2-Ba-N1	61.27	61.30	0.03
				N2-Ba-O6	143.21	143.20	0.01
				N1-Ba-O6	102.39	102.40	0.01
Complex (4)							
Na-N1	2.552	2.554	0.002	O1-Na-O2	99.56	99.50	0.06
Na-N2	2.565	2.566	0.001	O1-Na-O3	86.59	86.60	0.01
Na-O1	2.412	2.412	0.000	O1-Na-O4	174.34	174.30	0.04
Na-O2	2.393	2.393	0.000	O1-Na-O4A	90.18	9023	0.05
Na-O3	2.505	2.505	0.000	O1-Na-N1	69.82	69.82	0.00
Na-O4	2.628	2.629	0.001	O1-Na-N2	109.13	109.10	0.03
Na-O4A	2.442	2.443	0.001	O2-Na-O3	164.10	164.13	0.03
Na-NaA	3.429	3.430	0.001	O2-Na-O4	78.26	78.31	0.05
				O2-Na-O4A	88.56	88.57	0.01
				O2-Na-N1	70.69	70.70	0.01
				O2-Na-N2	121.68	121.71	0.03
				O3-Na-O4	96.86	96.87	0.01
				O3-Na-N1	125.19	125.17	0.02
				O3-Na-N2	68.93	68.91	0.02
				O3-Na-O4A	76.69	76.72	0.03
				O4-Na-O4A	94.96	95.00	0.03
				O4-Na-N1	68.25	68.25	0.00
				O4-Na-N2	68.25	68.25	0.00
				N1-Na-N2	73.38	73.37	0.01
				N1-Na-O4A	147.48	147.51	0.03
				N2-Na-O4A	138.85	138.87	0.02
Complex(5)							
Sr-O1	2.617	2.618	0.001	O3-Sr-O2	151.28	158.70	7.42
Sr-O2	2.611	2.610	0.001	O3-Sr-O1	73.24	77.20	3.96
Sr-O3	2.506	2.505	0.001	O2-Sr-O1	113.98	114.30	0.32
Sr-O4	2.618	2.618	0.000	O3-Sr-O4	87.91	88.20	0.29
Sr-O1W	2.701	2.702	0.001	O2-Sr-O4	77.89	73.50	4.39
Sr-O2W	2.699	2.699	0.000	O1-Sr-O4	158.36	151.60	6.76
Sr-O2WA	2.726	2.699	0.000	O3-Sr-N1	102.34	113.60	11.26
Sr-N1	2.835	2.835	0.000	O4-Sr-N1	113.34	102.50	10.84
Sr-N2	2.849	2.849	0.000	O3-Sr-N2	61.38	64.40	3.02

Bond distances (Å)	Theoretical	Experimental	Dev.	Bond angles (°)	Theoretical	Experimental	Dev.
				O2-Sr-N2	89.92	97.10	7.18
				O1-Sr-N2	96.80	90.2	6.60
				O4-Sr-N2	64.13	61.60	2.53
				O3-Sr-O2W	106.07	106.40	0.33
				O2-Sr-O2W	70.58	77.10	6.52
				O1-Sr-O2W	129.35	140.50	11.15
				O4-Sr-O2W	70.80	67.90	2.90
				O2-Sr-O1W	79.52	65.40	14.12
				O1-Sr-O1W	65.11	79.80	14.69
				O4-Sr-O1W	136.42	127.40	9.02
				O3'-Sr-O1W	126.14	127.90	1.76
				O2W-Sr-O1W	70.75	71.00	0.25
				O2W-Sr-N2	128.11	128.40	0.29
				O1W-Sr-N2	152.06	152.40	0.34
				N1-Sr-N2	64.77	65.00	0.23
				O2W-Sr-N1	139.03	139.30	0.27
				O1W-Sr-N1	87.52	87.80	0.28
Complex (**6**)							
Ba-O1	2.736	2.738	0.002	O4-Ba-O1	162.70	162.90	0.20
Ba-O2	2.763	2.761	0.002	O4-Ba-O3	92.45	92.70	0.25
Ba-O3	2.756	2.757	0.001	O1-Ba-O3	79.69	80.00	0.31
Ba-O4	2.658	2.657	0.001	O4-Ba-O2	84.25	84.60	0.35
Ba-O1W	2.880	2.880	0.000	O1-Ba-O2	94.22	94.50	0.28
Ba-O2W	2.988	2.990	-0.002	O3-Ba-O2	148.19	148.40	0.21
Ba-N1	3.009	3.002	0.007	O4-Ba-N1	107.19	107.50	0.31
Ba-N2	3.010	3.004	0.006	O1-Ba-N1	58.19	58.40	0.21
				O3-Ba-N1	91.90	92.00	0.01
				O2-Ba-N1	59.43	59.60	0.17
				O4-Ba-N2	59.95	61.10	1.15
				O1-Ba-N2	103.03	103.20	0.17
				O3-Ba-N2	58.68	58.80	0.12
				O2-Ba-N2	93.2	93.40	0.20
				N1-Ba-N2	61.84	62.00	0.16
				O4-Ba-O1W	126.17	126.50	0.33
				O1-Ba-O1W	66.29	66.50	0.23
				O3-Ba-O1W	72.53	72.80	0.27
				O2-Ba-O1W	133.50	133.80	0.30
				O1-Ba-O2W	73.43	73.70	0.27
				O3-Ba-O2W	131.18	131.50	0.32

Bond distances (Å)	Theoretical	Experimental	Dev.	Bond angles (°)	Theoretical	Experimental	Dev.
				O2-Ba-O2W	74.74	75.00	0.26
				O2W-Ba-O1WA	59.36	59.50	0.14
				N1-Ba-O1WA	137.03	137.20	0.17
				N1-Ba-O1W	124.19	124.19	0.00
				N1-Ba-O2W	106.68	106.90	0.22
				N2-Ba-O1WA	123.73	124.00	0.27
				N2-Ba-O1W	131.21	131.50	0.29
				N2-Ba-O2W	166.91	167.00	0.10
				N2-Ba-O2WA	100.84	100.87	0.03

Table S1.
*Comparison of selected experimental and calculated geometric parameters bond lengths (Å) and bond angles (°) for complexes (**1–6**).*

Author details

Rakesh Kumar[1]* and Sangeeta Obrai[2]

1 Department of Chemistry, MCM DAV College, Kangra, Himachal Pradesh, India

2 Department of Chemistry, Dr. B.R. Ambedkar National Institute of Technology, Jalandhar, Punjab, India

*Address all correspondence to: rakesh_nitj@yahoo.co.in

References

[1] Lehn JM. Structure and bonding. Alkali Metal Complexes with Organic Ligands. In: Dunitz JD, Hemmirch P, Ibers JA, Jorgensen CK, Neilands JB, Reinan D, Williams RJP, editors. Vol. 16, 7. New York: Springer-Verlag; 1973

[2] Tümmler B, Maas G, Weber E, Wehner W, Vögtle F. Noncyclic crown-type polyethers, pyridinophane cryptands, and their alkali metal ion complexes: Synthesis, complex stability, and kinetics. Journal of the American Chemical Society. 1977;**99**:4688

[3] Cram DJ, Cram JM. Design of complexes between synthetic hosts and organic guests. Accounts of Chemical Research. 1978;**11**:8

[4] Cram DJ. Preorganization - from solvents to spherands. Angewandte Chemie (International Ed. in English). 1986;**25**:1039

[5] Poonia NS, Bajaj AV. Coordination chemistry of alkali and alkaline earth cations. Chemical Reviews. 1979;**79**:145

[6] Bajaj AV, Poonia NS. Comprehensive coordination chemistry of alkali and alkaline earth cations with macrocyclic multidentates: Latest position. Coordination Chemistry Reviews. 1988;**87**:55

[7] Vögtle F, Müller WM, Buhleier E, Weber W. Synthesis and selectivity of novel four armed noncyclic neutral ligands. Chemische Berichte. 1979; **112**:899

[8] Vögtle F, Weber E. Multidentate acyclic neutral ligands and their complexation. Angewandte Chemie (International Ed. in English). 1979; **18**:753

[9] Vögtle F, Weber E. Crystalline 1: 1 alkali metal complexes of noncyclic neutral ligands. Tetrahedron Letters. 1975;**16**:2415

[10] Hay BP, Rustad JR. Structural criteria for the rational design of selective ligands: Extension of the MM3 force field to aliphatic ether complexes of the alkali and alkaline earth cations. Journal of the American Chemical Society. 1994;**116**:6316

[11] Wipff G, Weiner P, Kollman P. A molecular-mechanics study of 18-crown-6 and its alkali complexes: An analysis of structural flexibility, ligand specificity, and the macrocyclic effect. Journal of the American Chemical Society. 1982;**104**:3249

[12] Grootenhuis PDJ, Kollman PA. Molecular mechanics and dynamics studies of crown ether-cation interactions: Free energy calculations on the cation selectivity of dibenzo-18-crown-6 and dibenzo-30-crown-10. Journal of the American Chemical Society. 1989;**111**:2152

[13] Pretsch E, Badertscher M, Welti M, Maruizumi T, Morf WE, Simon W. Design features of ionophores for ion selective electrodes. Pure and Applied Chemistry. 1988;**60**:567

[14] Badertscher M, Welti M, Pretsch E, Maruizumi T, Ha T–K, Simon W. Combined application of pair potentials and the MM2 force field for the modeling of ionophores. Journal of Computational Chemistry. 1990;**11**:819

[15] Varnek AA, Glebov AS, Petrukhin OM, Kolycheva NV, Ozerov RP. Selectivity of reactions of 18-crown-6 complexation with alkali-metals. Koordinatsionnaya Khimiya (Russian Journal of Coordination Chemistry). 1989;**15**:600

[16] Burns JH, Kessler RM. Structural and molecular mechanics studies of bis(dibutylphosphato)aquastrontium-18crown-6 and analogous

alkaline-earth-metal complexes. Inorganic Chemistry. 1987;**26**:1370

[17] Wipff G, Kollman P. Molecular mechanical calculations on a macrocyclic receptor: The 222 cryptand and its alkali complexes. Nouveau Journal de Chimie. 1985;**9**:457

[18] Damu KV, Hancock RD, wade PW, Boeyens JCA, Billing DG, Dobson SM. Control of metal-ion size-based selectivity through the structure of the oxygen-donor pendant groups on lariat ethers. A crystallographic and thermodynamic study. Journal of the Chemical Society, Dalton Transactions. 1991:293

[19] Kollman P, Wipff G, Singh UC. Molecular mechanical studies of inclusion of alkali cations into anisole spherands. Journal of the American Chemical Society. 1985;**107**:2212

[20] Maye PV, Venanzi CA. Host-guest preorganization and complementarity: A molecular mechanics and molecular dynamics study of cation complexes of a cyclic urea-anisole spherand. Journal of Computational Chemistry. 1991;**12**:994

[21] Lifson S, Felder CS, Shanzer A. Computational design of polylactone macrocyclic ionophores. Journal of the American Chemical Society. 1983;**105**:3866

[22] Lifson S, Felder CS, Shanzer A. Design, synthesis, and conformation of ion carriers. Bopolymers. 1983;**22**:409

[23] Lifson S, Felder CS, Shanzer A. Internal and external alkali ion complexes of enniatin B: An empirical force field analysis. Biochemistry. 1984;**23**:2577

[24] Lifson S, Felder CS, Shanzer A. Enniatin B and valinomycin as ion carriers: An empirical force field analysis. Journal of Biomolecular Structure & Dynamics. 1984;**2**:641

[25] Kumar R, Obrai S, Kaur A, Hundal G, Meehnian H, Jana AK. Synthesis, crystal structure determination and antimicrobial studies of copper(II) picrate complexes with N, N, N‴-tetrakis(2-hydroxyethyl/propyl) ethylenediamine and tris (2-hydroxyethyl)amine. Polyhedron. 2013;**56**:55

[26] Kumar R, Obrai S, Kaur A, Hundal MS, Meehnian H, Jana AK. Synthesis, crystal structure investigation, DFT analyses and antimicrobial studies of silver(I) complexes with N, N, N‴-tetrakis (2-hydroxyethyl/propyl) ethylenediamine and tris(2-hydroxyethyl)amine. New Journal of Chemistry. 2014;**38**:1186

[27] Kumar R, Obrai S, Jassal AK, Hundal MS. Supramolecular architectures of N, N,N″,N″-tetrakis-(2-hydroxyethyl) ethylenediamine and tris(2-hydroxyethyl)amine with La(III) picrate. RSC Advances. 2014;**4**:59248

[28] Akhtar MN, Harrison WTA, Shahid M, Khan IU, Ejaz JI. Synthesis, crystal structure and biological activity of a cobalt(II) complex of N,N,N″-tetrakis (2-hydroxypropyl) ethylenediamine. Transition Metal Chemistry. 2016;**41**: 325-330

[29] Frisch MJ, Trucks GW, Schlegel HB, Scuseria GE, Robb MA, Cheeseman JR, et al. Gaussian 03, (Revision B.04). Wallingford, Connecticut: Gaussian, Inc.; 2004

[30] Becke AD. Density-functional thermochemistry. III. The role of exact exchange. The Journal of Chemical Physics. 1993;**98**:5648

[31] Lee C, Yang W, Parr RG. Development of the Colle-Salvetti correlation-energy formula into a functional of the electron density. Physical Review B. 1988;**37**:785

[32] Patersson GA, Al-Laham MA. A complete basis set model chemistry. II.

Open-shell systems and the total energies of the first-row atoms. The Journal of Chemical Physics. 1991;**94**: 6081

[33] Hay PJ, Wadt WR. Ab initio effective core potentials for molecular calculations. Potentials for the transition metal atoms Sc to Hg. The Journal of Chemical Physics. 1985;**82**:270

[34] Wadt WR, Hay PJ. Ab initio effective core potentials for molecular calculations. Potentials for main group elements Na to Bi. The Journal of Chemical Physics. 1985;**82**:284

[35] Hundal G, Hundal MS, Obrai S, Poonia NS, Kumar S. Metal complexes of tetrapodal ligands: synthesis, spectroscopic and thermal studies, and X-ray crystal structure studies of Na(I), Ca(II), Sr(II), and Ba(II) complexes of tetrapodal ligands N,N,N′,N′-tetrakis(2-hydroxypropyl)ethylenediamine and N,N,N′,N′-tetrakis(2-hydroxyethyl) ethylenediamine. Inorganic Chemistry. 2002;**41**:2077

[36] Mariappan G, Sundaraganesan N. Spectral and structural studies of the anti-cancer drug Flutamide by density functional theoretical method. Spectrochimica Acta Part A. 2014; **117**:604

[37] Fleming I. Frontier Orbitals and Organic Chemical Reactions. New York: John Wiley and Sons; 1976

[38] Balachandran V, Lalitha S, Rajeswari S, Rastogi VK. Theoretical investigations on the molecular structure, vibrational spectra, thermodynamics, HOMO-LUMO, NBO analyses and paramagnetic susceptibility properties of p-(p-hydroxyphenoxy)benzoic acid. Spectrochimica Acta Part A. 2014; **121**:575

[39] Raif K, Sinem O, Azizoglu A. Experimental and computational study on [2,6-bis(3,5-dimethyl-N-pyrazolyl) pyridine]-(dithiocyanato)mercury(II). Polyhedron. 2007;**26**:5069

[40] Uesugi Y, Mizuno M, Shimojima A, Takahashi H. Resonance Raman and ab initio MO calculation studies of the structures and vibrational assignments of the T1 state and the anion radical of coumarin and its isotopically substituted analogues. The Journal of Physical Chemistry. 1997;**101**:268

Section 3

Ionic Liquid

Chapter 3

Theoretical Study of the Structure and Property of Ionic Liquids as Corrosion Inhibitor

Guocai Tian and Weizhong Zhou

Abstract

Three sets of ionic liquids such as 1-alkyl-3-methylimidazole chloride [$C_n$mim] Cl, 1-alkyl-3-methylimidazolium acetate [$C_n$mim]Ac and 1-octyl-3-methylimidazole salt [Omim]Y (n = 2, 4, 6, 8, and Y = Cl, BF_4, HSO_4, Ac and TFO) were used as corrosion inhibitor medium for corrosion protection of carbon steel. Electronic structures and reactivity of these ionic liquids, surface energy and electronic structures of the iron surface were systematically analyzed by density functional theory. By increasing the alkyl chain length of the [$C_n$mim]Cl and [$C_n$mim] Ac systems, the lowest unoccupied molecular orbital energy (E_{LUMO}), the highest occupied molecular orbital energy (E_{HOMO}), the softness (S) and polarizability (α) increased gradually, whereas electronegativity (χ), energy gap (ΔE), hardness (η), dipole moment (μ)and electrophilic index (ω) gradually decreased. For the [Omim] Y system, the structure parameters of ionic liquids are quite different, and only the polarizability (α) decreases gradually by increasing the length of the alkyl chain. The results show that inhibition is mainly [$C_n$mim]$^+$ cations of the [$C_n$mim]Cl system, and the order of inhibition efficiency follows as [C_2mim]Cl < [C_4mim]Cl < [C_6mim]Cl < [C_8mim]Cl. Both [$C_n$mim]$^+$ cations and the Ac^- anion have inhibition effect for the [Xmim]Ac system, and the order of inhibition efficiency is [C_8mim]Ac > [C_6mim]Ac > [C_4mim]Ac > [C_2mim]Ac. For the [Omim]Y system, [Xmim]$^+$ cations and anions (BF_4^-, HSO_4^-, Ac^-, TfO^-) have inhibition effect, and the order of inhibition efficiency is [Omim]TfO > [Omim]Ac > [Omim] HSO_4 > [Omim]BF_4 > [Omim]Cl.

Keywords: inhibitor, carbon steel, density functional theory, active site, global and local activity parameters, geometry structure

1. Introduction

Metal corrosion is a ubiquitous phenomenon in the national economy, national defense construction, science and technology and other relative fields [1–3]. It has tremendous negative effects and need to be treated. Corrosion of metal and its alloys in various fields in each country has caused serious problems in different fields and caused a series of major losses. Therefore, we must adopt effective corrosion protection measures. As latest reported from the National Association of Anticorrosive Engineers, the world's annual corrosion cost can decrease by 15–35% (US $ 375–75 billion) by adopting appropriate anticorrosion technology [1–3].

 IntechOpen

Due to its low cost and excellent physical, chemical and mechanical properties, iron and its alloys are widely used in many fields, and most of them are in contact with corrosive environments, such as tanks, pipelines, boilers, oil and gas production units and refineries [1]. Therefore, it is necessary to take certain protective measures, and the addition of an inhibitor can significantly reduce the corrosion rate of iron and ferroalloy, and the amount of inhibitor is small, which has become an economic and effective method [2]. Traditional inhibitors, such as chromate, nitrite, silicate and molybdate, can react with metals to constitute a passive film or a dense metal salt protective film on the metal surface, preventing the corrosion of metals and alloys [3]. However, this kind of inhibitor is usually used in large amount and high cost. In addition, when the amount of inhibitor is insufficient, it will lead to serious local corrosion of metals and alloys. In addition, these inhibitors have certain toxicity and poor biodegradation ability. When they are accustomed, there are serious environmental and ecological problems and they are increasingly restricted. Therefore, it is very important to develop environment-friendly inhibitors. As an environment-friendly inhibitor, imidazolines, amides and other inhibitors have the characteristics of high efficiency, low toxicity and easy biodegradation, which are commonly used in the corrosion inhibition of iron and alloy [1–3].

In recent years, ionic liquids (ILs) as a new green solvent, which is formed by organic cations with various inorganic or organic anions, have attracted more and more attention [4–6]. Due to the heterocyclic structure, heteroatoms (such as N, O and S) and multiple bonds in ionic liquid molecules, they have become effective inhibitors and are commonly used in the field of corrosion protection [6–9]. The ionic liquid common inhibitors are imidazole ionic liquid, pyridine ionic liquid and quaternary ammonium ionic liquid. Compared with traditional corrosion inhibitors, ionic liquid corrosion inhibitors are always renewable, inexpensive, ecologically acceptable, readily available and environmentally friendly and biodegradable. Various ionic liquids such as imidazoline, pyridine, ammonium, pyridazine and benzimidazole were widely used as corrosion inhibitors for carbon steel. The experimental methods such as surface analysis techniques, electrochemical tests and electrochemical impedance spectroscopy are used extensively to study the performance of ionic liquid inhibitors [9–25].

However, experimental research on ionic liquid inhibitors mainly focuses on the evaluation of their corrosion inhibition performance. The research on the corrosion inhibition mechanism of ionic liquid inhibitors is less, which can not reveal some details of the corrosion inhibition process, such as the quantitative description of the molecular metal atom combination of inhibitors, the determination of the functional groups of inhibitors and the information on the process of ion adsorption and film formation of inhibitors [9–11]. It is impossible to guide the development of ionic liquid inhibitors by obtaining and diffusing the inhibitors on the metal surface. Therefore, it is necessary to study the mechanism of ionic liquid inhibitors by theoretical calculation, so as to provide theoretical guidance for the development and application of new and efficient ionic liquid inhibitors.

With the rapid improvement and advancement of computer science and technology, and the gradual improvement of related theories, quantum chemical calculation as an important theoretical calculation method has been proved to be very effective in determining the electronic structure of molecules and clarifying the reaction activity of molecules. The inhibition mechanism and performance of inhibitors are often the same as the electronic structure and reaction activity of inhibitors. Therefore, it is very important to study the electronic structure and reaction activity of ionic liquid inhibitors for the design of new ionic liquid inhibitors [26]. Murulana et al. [27] studied 1-octyl-3-methylimidazolium bis (trifluoromethyl-sulfonyl) imide ([Hmim]NTF_2), 1-propyl-3-methylimidazolium-bis(trifluoromethyl-sulfonyl) imide ([Pmim]NTF_2), 1-propyl-2,3-dimethylimidazolium bis(trifluoromethyl-sulfonyl) imide [PDmim]NTF_2 and 1-butyl-3-methylimidazolium bis(trifluoromethyl-sulfonyl) imide ([Bmim]NTF_2). The results showed that [PDmim]NTF_2 has the strongest capability of electron donor, and [Pmim]NTF_2 has capability of electron acceptor. The active centers are located in the ring of imidazole cation and the N and O atoms of anions. Sasikumar et al. [28] studied the behavior of 1-decyl-3-methylimidazolium tetrafluoride ([Dmim]BF_4), 1-ethyl-3-methylimidazolium boron tetrafluoride ([Emim]BF_4) and 1-butyl-2,3-dimethylimidazolium tetrafluoride ([BDmim]BF_4) inhibition on low carbon steel by density functional theory in hydrochloric acid solution. The results showed that the variations of the E_{HOMO}, hardness (η) and dipole moment (μ) are in agreement with the electrochemical experiments. The order of inhibition efficiency is [Dmim]BF_4 > [BDmim]BF_4 > [Emim]BF_4, which agrees well with experiments. The corrosion inhibition behavior of 1-hexyl-3-methylimidazolium iodide ([Hmim]I), 1-hexyl-3-methylimidazolium phosphorus hexafluoride ([Hmim]PF_6), 1-hexyl-3-methylimidazolium trifluoromethanesulfonate ([Hmim]TFO) and 1-hexyl-3-methylimidazolium boron tetrafluoride ([Hmim]BF_4) on carbon steel was studied by Mashuga et al. [29] with density functional theory. It was found that the variation order of E_{HOMO}, hardness (η), softness (S) and the dipole moment (μ) is [Hmim]PF_6 < [Hmim]BF_4 < [Hmim]I < [Hmim]TFO. The results agree well with the experimental inhibition efficiency measured by electrochemical methods. The inhibition performance of two 1-ethyl-3-methylimidazolium-based ionic liquids formed with ethyl sulfate ([Emim]$EtSO_4$) and acetate ([Emim]Ac) and three 1-butyl-3-methylimidazolium-based ionic liquids with acetate ([Bmim]Ac), thiocyanate ([Bmim]SCN) and dichloride ([Bmim]DCA) on carbon steel in acid medium was studied with density functional theory by Yesudass et al. [30]. The results show that the inhibition efficiency follows [Emim]Ac < [Bmim]SCN < [Bmim]Ac < [Bmim]DCA < [Emim]$EtSO_4$, which is consistent with the experimental results.

As mentioned above, although some progress has been made in the study of the corrosion mechanism and properties of low carbon steel by ionic liquid corrosion inhibitors in the corrosion medium, there are still some problems. Experimental methods (weight-loss method, electrochemical experiment method and surface morphology analysis method) are used to study the ionic liquid, mainly focusing on the evaluation of the inhibition performance and efficiency of ionic liquid, but it is difficult to know the inhibition mechanism of ionic liquid on the iron surface. However, the structure parameters calculated by quantum chemistry are not systematically analyzed. For example, for the imidazole system with different cations, how do the structure parameters and property change with the increase of the cationic alkyl chain length? For the system with the same cations and different anions, how do the structure parameters change? There is no systematic analysis of the properties of different surfaces of iron, but it simply selects a surface as the adsorption surface of ionic liquid. At the same time, different surface of iron may have different electric charge and different electric charge. There are some relations between iron and HOMO and LUMO of ionic liquid molecules, but there are few related researches.

Molecular simulation is only a simple calculation of adsorption energy between the metal surface and ionic liquid inhibitors, but there are few studies on the quantitative description of the molecular metal atom combination of ionic liquid, the determination of the functional groups of ionic liquid, and the acquisition of the process information of adsorption and film formation, which cannot guide the development of ionic liquid corrosion inhibitors. In this paper, the equilibrium geometry, frontier orbit, global and local reactivity of [$C_n$mim]Cl system, [$C_n$mim]Ac system and

[Omim]Y (n = 2, 4, 6, 8, Y = Cl, BF_4, HSO_4, Ac and TFO) are studied by quantum chemical calculation. The active region, inhibition efficiency of possible interaction between ionic liquid molecules and iron surface are preliminaries analyzed.

2. Computational method

The B3LYP method of density functional theory in Gaussian 09 software package [31] is utilized to optimize the geometric configuration and analyze the frequency of ionic liquid. The 6-311++G(d, p) basis set was used in this paper, which is same as the literature [26–28]. The geometry optimization and frequency analysis of the studied ionic liquid molecules were performed, and the stable geometry was obtained. There are seven interaction sites of 1-alkyl-3-methylimidazolium cations to interact with anions, as shown in **Figure 1**, where 1–5 represent the five positions of anions in the plane of imidazole ring, and 6 and 7 are respectively located on both sides perpendicular to the plane of imidazole ring. As confirmed by many authors, position 2 is the most active site and the anions are always located in this position, so we just discuss the anions' contact with cations at position 2. The E_{HOMO} and E_{LUMO}, energy gap (ΔE), electronegativity (χ), the softness (S), dipole moment (μ), polarizability (α), electrophilic index (ω), hardness (η), Fukui index and other structural and property parameters were calculated. Distribution of Fukui index was drawn by Multiwfn software [32, 33].

According to the frontier molecular orbital theory, the molecules' reactivity is related to their lowest unoccupied molecular orbits (LUMO) and the highest occupied molecular orbital (HOMO). In general, the level of the E_{HOMO} value determines the ability to deliver electrons to the outside. The level of the E_{LUMO} value represents the ability to obtain electrons. Therefore, the higher the E_{HOMO} value, the lower the E_{LUMO} value can reflect the reactivity of a molecule. [34] As for the adsorption of a molecule on the carbon steel surface, a higher E_{HOMO} value will make them have a stronger electronic supply capacity and can form an adsorption with the d orbital on the surface of iron. A lower E_{LUMO} value will make them have a stronger electronic acceptance capacity, which becomes more easily accepted electrons from the surface of iron and form an antibond orbital to promote the adsorption. Therefore, the energy gap difference ΔE = E_{HOMO}-E_{LUMO} can be used to assess

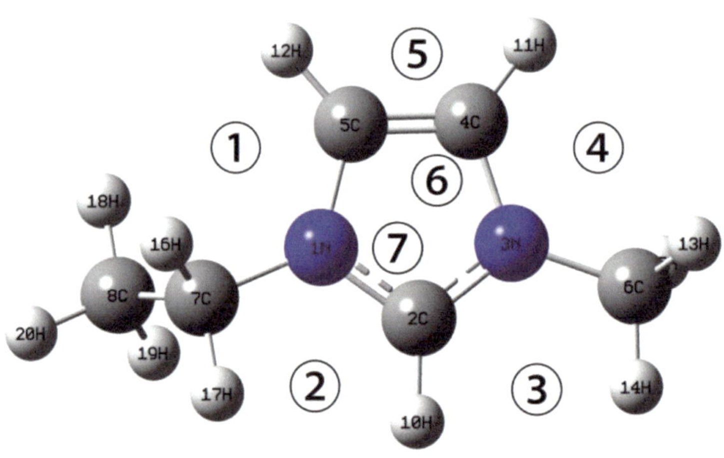

Figure 1.
Anions can form ionic liquids with [Emim]⁺ in several different sites.

the bonding ability of a molecule with iron. The smaller Δ_E is, the stronger bonding ability with iron is, and the easier adsorption on the iron surface [35].

In the concept density functional theory, according to Koopmans theorem [36], ionization potential I under vacuum condition can be approximated to the negative value of E_{HOMO} $I = -E_{HOMO}$. Similarly, the electron affinity a is approximately negative for E_{LUMO} $A = -E_{LUMO}$. The dipole moment (μ), electronegativity (μ), softness (S) and polarizability (α) of a molecule are generally referred to as the global reactivity of a molecule [37, 38]. Molecular polarity is always described by dipole moment (μ), which is defined as the product of atomic distance R and atomic charge q as $\mu = qR$ [39]. Generally, the higher μ will make the molecules more easily adsorbed on the surface of carbon steel [40], and the efficiency of the molecular inhibitor will increase accordingly. However, the dipole moment reflects the global polarity of molecules rather than the polarity of the bond in the molecule. According to the concept density functional theory, the electronegativity (χ) of the system with n electrons is defined as follows when the external field $\nu(r)$ is fixed [38]:

$$\chi = -\left(\frac{\partial E}{\partial N}\right)_{\nu(r)} \tag{1}$$

Hardness (η) is defined as the second derivative of the system energy E to N [41]

$$\eta = \frac{1}{2}\left(\frac{\partial^2 E}{\partial N^2}\right)_{\nu(r)}. \tag{2}$$

Using the finite difference approximation, the global hardness (η) and electronegativity (χ) are related to ionization energy I and affinity energy A as follows [42]:

$$\chi = \frac{I + A}{2}, \tag{3}$$

$$\eta = \frac{I - A}{2} \tag{4}$$

According to Koopmans theorem [36], the electronegativity (χ) and global hardness (η) can be calculated from E_{HOMO} and E_{LUMO} as

$$\chi = -\frac{E_{HOMO} + E_{LUMO}}{2}, \tag{5}$$

$$\eta = \frac{E_{LUMO} - E_{HOMO}}{2}. \tag{6}$$

Global softness (s) is usually defined as the reciprocal of global hardness [37].

$$S = \frac{1}{\eta}. \tag{7}$$

Electronegativity (χ) is a scale of the ability of atoms in molecules to attract electrons. The larger its value is, the easier it is to attract electrons [38] reflecting a better inhibitor effect. The smaller the global hardness (η) or the larger the global softness (S) of the molecule means that the stronger the interaction between metal surface and a molecule [43], and a higher corrosion inhibition efficiency. Parr introduced the concept of electrophilic index (ω), which is defined as following [37]:

$$\omega = \frac{\chi^2}{2\eta}. \tag{8}$$

According to the definition, this parameter is a measure of the capability of electron acceptors. The larger the value is, the stronger the capability of electron acceptor is. As an important parameter of global reaction activity, the molecular polarizability (α) is the average value obtained by calculation, and its relationship is as follows [44]:

$$\alpha = \frac{1}{3}\left(\alpha_{xx} + \alpha_{yy} + \alpha_{zz}\right). \tag{9}$$

where, α_{xx}, α_{yy} and α_{zz} are molecular polarizability in x, y and z directions, respectively. The higher the molecular polarizability (α) is, the easier adsorption on the metal surface is, and the higher corrosion inhibition performance is [44].

The local reactivity of ionic liquid inhibitors was evaluated by their Fukui index. Nucleophilic and electrophilic behavior of the molecule was studied by analyzing its Fukui index distribution. Fukui function is an important concept in density functional theory, which is commonly used to predict the active sites of a molecule [45]. In the case of certain outfield $v(r)$, the Fukui function is defined as follows [46]:

$$f(r) = \left[\frac{\partial \rho(r)}{\partial N}\right]_{\nu(r)}. \tag{10}$$

Within the finite difference approximation, the nucleophilic attack can be expressed as $f^{+}(r) = \rho_{N+1}(r) - \rho_N(r)$, and electrophilic attack is $f^{-}(r) = \rho_N(r) - \rho_{N-1}(r)$ [29], where $\rho_{N+1}(r)$、 $\rho_N(r)$、 $\rho_{N-1}(r)$ are the charge density of the atom in the molecule with one unit negative charge, uncharged and one unit positive charge, respectively.

3. Results and discussion

3.1 1-Alkyl-3-methylimidazole chloride [C$_n$mim]Cl

In this section, we will analyze the equilibrium geometry, frontier orbital distribution, global and local reactivity, the change rule of the reactive sites and structure parameters of ionic liquids with the increase of the cationic alkyl chain length for the [C$_n$mim]Cl system. The comparison of the corrosion inhibition performance was achieved for the ionic liquids formed with the same anion and different cations.

3.1.1 Equilibrium geometry structures, HOMO and LUMO

Figure 2 shows the equilibrium geometry, HOMO and LUMO distribution of [C$_n$mim]Cl (n = 2, 4, 6, 8) system obtained by B3LYP/6-311++G(d, p) method. It can be seen from **Figure 2** that the equilibrium geometry HOMO and LUMO profiles of these four ionic liquids are very similar. The anions are all at the C2 atom of the imidazole ring. HOMO is distributed on anions, but not on cations, which indicates that Cl^{-} is easy to form a coordination bond with d-orbital on an iron surface, while imidazole ring is not easy to provide electrons and interact with the iron surface. LUMO is mainly distributed on the imidazole ring, and other C atoms and anions of cation also have little distribution, which indicates that the imidazole

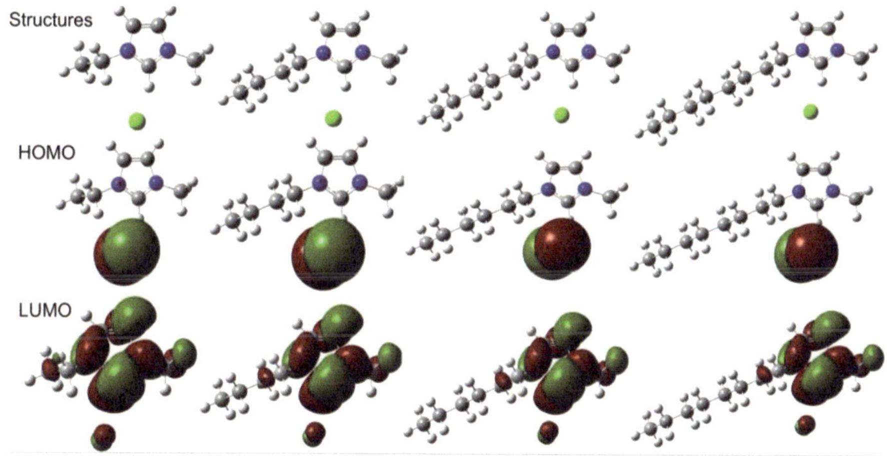

Figure 2.
The equilibrium geometry structures, HOMO and LUMO isosurfaces of [$C_n$mim]Cl (n = 2,4,6,8) with B3LYP/6-311++G(d,p) method. From left to right, it is [C_2mim]Cl, [C_4mim]Cl, [C_6mim]Cl and [C_8mim] Cl, respectively.

ILs	[C_2mim]Cl	[C_4mim]Cl	[C_6mim]Cl	[C_8mim]Cl
E_{HOMO} (eV)	−5.2672	−5.2641	−5.2623	−5.2608
E_{LUMO} (eV)	−1.1890	−1.1824	−1.1820	−1.1819
ΔE (eV)	4.0782	4.0817	4.0903	4.0789

Table 1.
E_{HOMO}, E_{LUMO} *and* ΔE *of [$C_n$mim]Cl (n = 2, 4, 6, 8) with B3LYP/6-311++G (d, p) method.*

ring of cation is easy to accept electrons from the iron surface and form a feedback bond. According to the distribution of HOMO and LUMO, when the [Xmim]Cl ionic liquid adsorbs on the surface of carbon steel, the imidazole ring of the ionic liquid will interact with the surface of carbon steel and lay parallel to the surface.

The E_{HOMO}, the E_{LUMO} and the energy gap difference (ΔE) of four ionic liquid molecules in [$C_n$mim]Cl system are shown in **Table 1**. It can be seen from **Table 1** that by increasing the length of the alkyl chain, the E_{HOMO} becomes larger and larger, indicating the increased capability of electron donor of the molecule. By increasing the length of the alkyl chain, E_{LUMO} also has an increasing trend, which shows that the capability of electron acceptor will weaken. However, the smaller (ΔE) is, the better the activity of the molecule is, the easier absorption between the carbon steel surface and molecules, and the higher the inhibition efficiency is. The sequence of inhibition efficiency of these four ionic liquids should be [C_2mim]Cl < [C_4mim]Cl < [C_6mim]Cl < [C_8mim]Cl. The [C_8mim]Cl has the highest inhibition efficiency and the best inhibition performance, which agrees well with the experimental measurement [29].

3.1.2 Global activity parameters

Table 2 shows the global activity parameters of [$C_n$mim]Cl system obtained by B3LYP/6-311++G(d, p) method. It can be seen from **Table 2** that by increasing the length of the alkyl chain, the dipole moment (μ) decreases gradually, indicating that the increase in alkyl chain length will reduce the polarity of the whole molecule. Increasing the length of alkyl chain, the electronegativity (χ) is also gradually

ILs	μ/Debye	χ/eV	η/eV	S/eV^{-1}	ω/eV	α(a.u.)
[C_2mim]Cl	12.6974	3.2281	2.0391	0.4904	2.5552	117.8930
[C_4mim]Cl	12.3672	3.2232	2.0409	0.4900	2.5453	132.8403
[C_6mim]Cl	12.2509	3.2222	2.0402	0.4902	2.5445	158.3847
[C_8mim]Cl	12.2083	3.2213	2.0394	0.4903	2.5441	183.8247

Table 2.
Global activity parameters for [$C_n$mim]Cl (n = 2, 4, 6, 8) system with B3LYP/6-311++G(d,p) method.

reduced, which indicates that the ability of molecules to attract electrons is weaker and weaker, which is not conducive to accepting the electrons on the iron surface. The global hardness (η) of the molecules decreases, while the global softness (S) increases by increasing the chain length, which indicates that the interaction between the surface of carbon steel with the molecules becomes stronger and stronger, and the molecules tend to adsorb on the surface. The electrophilic index (ω) decreases by increasing the length of the alkyl chain. With the lowest unoccupied molecular orbital energy (E_{LUMO}), the ability of the molecule to accept electrons becomes weaker and weaker. Molecular polarizability (α) is an important quantum chemical parameter in the field of corrosion protection. From **Table 2**, the order of the magnitude of [C_8mim]Cl > [C_6mim]Cl > [C_4mim]Cl > [C_2mim]Cl can be concluded, indicating that [C_8mim]Cl is most easily adsorbed on the surface of iron and its corrosion inhibition efficiency is the highest. According to the analysis results in Section 3.1.1, with the increase of the alkyl chain length, the global activity parameters of [$C_n$mim]Cl system, such as global hardness (η), global softness (S) and polarizability (α), increases; therefore, the corrosion inhibition efficiency increases [29].

3.1.3 Local activity parameters

Figure 3 shows the Fukui index f^- distribution of four ionic liquid molecules in [$C_n$mim]Cl system, with the electrophilic attack index distribution on the left and the nucleophilic attack index f^+ distribution on the right. It can be seen from **Figure 3** that the electrophilic attack index of these four ionic liquid molecules is mainly distributed on the Cl atom with the HOMO.

It is easy to provide electrons when attacked by the dielectric since Cl atom has high electronegativity and high electron density around them. As the same of distribution of LUMO, the nucleophilic attack index of these four kinds of ionic liquid molecules is mainly distributed on the imidazole ring, especially 2C, 4C, 5C and 1N atoms (see **Figure 1** for the number of specific atom). The imidazole ring is

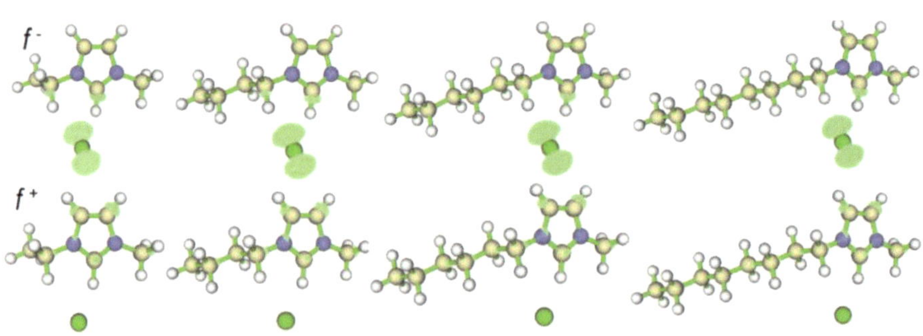

Figure 3.
Fukui indices isosurfaces of [$C_n$mim]Cl (n = 2, 4, 6, 8) system with B3LYP/6-311++G(d,p) method. The [C_2mim]Cl, [C_4mim]Cl, [C_6mim]Cl and [C_8mim]Cl are shown from left to right, respectively.

positively charged, and it is easier to accept electrons when attacked by the nucleophilic medium. Therefore, local reactive sites of $[C_nmim]Cl$ are Cl, 2C, 4C, 5C and 1N atoms (see **Figure 1** for the number of specific atom), which are easy to interact with iron surface and adsorb on iron surface.

3.2 1-Alkyl-3-methylimidazolium acetate $[C_nmim]Ac$ system

In this part, we will discuss how the change of chain length affects the change of structure and property of $[C_nmim]Ac$ (n = 2, 4, 6, 8) system with the same anion and different cations through the density functional theory. The equilibrium geometry, frontier orbital distribution, global reactivity and local reactivity, the change rule of reactive sites and structure parameters of ionic liquids by increasing the length of cationic alkyl chain and the comparison of corrosion inhibition performance are obtained and analyzed. In addition, the influence of different anions Cl^-, Ac^- on the reaction activity of ionic liquids was analyzed by comparing $[C_nmim]Ac$ (n = 2, 4, 6, 8) system with $[C_nmim]Cl$ system.

3.2.1 Equilibrium geometry and frontier orbital distribution

Figure 4 shows the equilibrium geometry, HOMO and LUMO frontier molecular orbital distribution of $[C_nmim]Ac$ (n = 2, 4, 6, 8) system obtained by B3LYP/6-311++G(d,p) method. It can be seen from **Figure 4** that the equilibrium geometry HOMO and LUMO distribution for these four ionic liquids is very similar. The anion Ac^- is located under the cation, and the optimized C—O bond is a double bond. HOMO is mainly distributed on the anion Ac^-, especially O atom, while C atom on the cation also has a small contribution, which shows that the unpaired O atom on the anion easily forms a coordination bond with the empty d-orbital on the iron surface and adsorbs on the iron surface. LUMO is mainly distributed on the imidazole ring, and there is a small amount of C of other cations and anions, which shows that the imidazole ring of cations easily accepts electrons from the iron

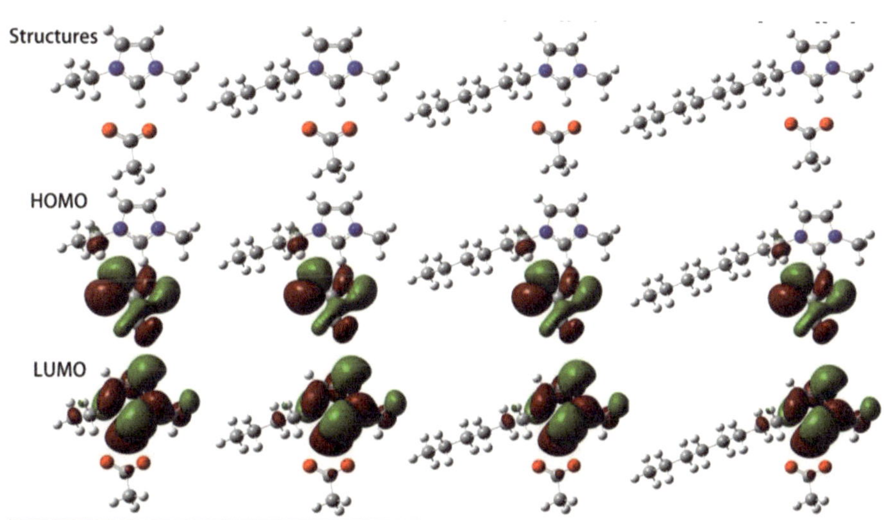

Figure 4.
Equilibrium geometry structures, HOMO and LUMO isosurfaces of $[C_nmim]Ac$ (n = 2, 4, 6, 8) system with B3LYP/6-311++G(d, p) method. From left to right, it is $[C_2mim]Ac$, $[C_4mim]Ac$, $[C_6mim]Ac$ and $[C_8mim]Ac$, respectively.

surface, forming a feedback bond, which makes the imidazole ring easy to be adsorbed on the iron surface.

The E_{HOMO}, E_{LUMO} and the energy gap ΔE of four [$C_n$mim]Ac (n = 2, 4, 6, 8) ionic liquid are shown in **Table 3**. It is shown in **Table 3** that E_{HOMO} and E_{LUMO} increase as that of [$C_n$mim]Cl (n = 2, 4, 6, 8) in **Table 2**, by increasing the length of alkyl chain, indicating that the capability of electron donator enhances and the capability of electron acceptor weakens. However, the smaller ΔE is, the better the activity of the molecule is, the easier it is to be adsorbed on the surface of iron and the better the inhibition effect is. The sequence of inhibition efficiency of the four ionic liquids should be [C_8mim]Ac > [C_6mim]Ac > [C_4mim]Ac > [C_2mim]Ac, and the inhibition efficiency of [C_8mim]Ac is the highest. The order obtained is the same as that of system analysis.

3.2.2 Global active parameters

Table 4 shows the global active parameters of [$C_n$mim]Ac system obtained by B3LYP/6-311++G(d,p) method. It can be seen from **Table 4** that by increasing the length of the alkyl chain, the dipole moment (μ) decreases gradually. Increasing the alkyl chain length will reduce the polarity of the whole molecule. The electronegativity (χ) also decreases gradually, which indicates that the capability of attracting electrons is weaker and weaker, and is not conducive to receiving electrons on the iron surface. This results are the same as those of [$C_n$mim]Cl system. As the alkyl chain length increases, the global hardness (η) and the global softness (S) of the molecule decrease, which indicates a stronger interaction between the carbon surface with the molecule and the molecule is more stably adsorbed on the iron surface. The electrophilic index (ω) decreases by increasing the alkyl chain length. With the increase of E_{LUMO}, the capability of electron acceptor of a molecule is weaker and weaker.

It can be seen from **Table 4** that the order of α is [C_8mim]Ac > [C_6mim]Ac > [C_4mim]Ac > [C_2mim]Ac, indicating that [C_8mim]Ac is most easily adsorbed on the iron surface. So, [C_8mim]Ac should have the best inhibition effect and the highest inhibition efficiency. Similar to the [$C_n$mim]Cl system, by increasing the length of alkyl chain, the values of dipole moment (μ), electronegativity (χ), global

ILs	[C_2mim]Ac	[C_4mim]Ac	[C_6mim]Ac	[C_8mim]Ac
E_{HOMO}	−5.4007	−5.3992	−5.3951	−5.3943
E_{LUMO}	−0.9120	−0.9009	−0.8979	−0.8985
ΔE	4.4887	4.4982	4.4972	4.4958

Table 3.
E_{HOMO}, E_{LUMO} *and* $\Delta E(eV)$ *of* [$C_n$mim]Ac (n = 2, 4, 6, 8) *with B3LYP/6-311++G(d, p) method.*

ILs	μ/Debye	χ/eV	η/eV	S/eV^{-1}	ω	α(a.u.)
[C_2mim]Ac	9.9354	3.1563	2.2444	0.4456	2.2194	120.9223
[C_4mim]Ac	9.6985	3.1500	2.2491	0.4446	2.2059	146.3850
[C_6mim]Ac	9.6243	3.1465	2.2486	0.4447	2.2015	171.8383
[C_8mim]Ac	9.5918	3.1464	2.2479	0.4449	2.2020	197.2063

Table 4.
Global activity parameters for [$C_n$mim]Ac (n = 2, 4, 6, 8) *system with B3LYP/6-311++G(d,p) method.*

hardness (η) and electrophilic index (ω) decrease, while the global activity parameters global hardness (η), global softness (S) and polarizability (α) increase. The order of polarization (α) is in agreement with that of corrosion inhibition efficiency.

3.2.3 Fukui index distribution

Figure 5 shows the Fukui index distribution of four ionic liquid [$C_n$mim]Ac systems. It can be seen from **Figure 5** that the electrophilic attack index of these four ionic liquid molecules is mainly distributed on the anion Ac^-, especially O atom, and the cation also has a small amount of distribution, indicating that the anion containing oxygen group is easy to provide the d-orbital combination of electron and iron surface space, and can stably be adsorbed on the iron surface. The nucleophilic attack index is mainly distributed in 2C, 4C and 5C atoms in imidazole ring (the number of atoms is shown in **Figure 1**) and O atom of anion, indicating that both cations and anions of [$C_n$mim]Ac system will interact with the surface of iron, making the adsorption of molecules on the surface of iron more stable. So the local reactive sites of [$C_n$mim]Ac are placed in O, 2C, 4C, 5C and 1N atoms. Local reactive sites are almost the same for [$C_n$mim]Ac and [$C_n$mim]Cl, except for Cl in [$C_n$mim]Cl.

3.3 1-Octyl-3-methylimidazole [Omim]Y system

In this section, we will discuss the structure and property of [Omim]Y (Y = Cl, BF_4, HSO_4, Ac, TFO) ionic liquids formed with the same cation and different anions. The equilibrium geometry, frontier orbital distribution, global reactivity and local reactivity, the change of reactive sites, structure parameters and corrosion inhibition performance of ionic liquids with different anions are obtained and discussed.

3.3.1 Equilibrium geometry and frontier orbital distribution

Figure 6 shows the equilibrium geometry, HOMO and LUMO distribution of [Omim]Y system obtained by B3LYP/6-311++G(d, p) method. It can be seen from **Figure 6** that the equilibrium geometry of the four ionic liquids is very similar, and the anions are all under the cations. LUMO is mainly distributed on the imidazole ring in the same way as [$C_n$mim]Cl system and [$C_n$mim]Ac system, which indicates that the imidazole ring of cation can easily accept electrons from the iron surface

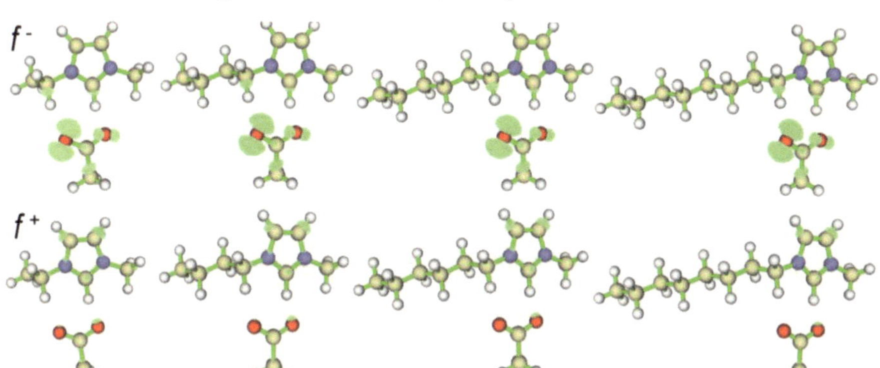

Figure 5.
Fukui index isosurfaces of [$C_n$mim]Ac (n = 2, 4, 6, 8) system with B3LYP/6-311++G(d,p) method. From left to right, it is [C_2mim]Ac, [C_4mim]Ac, [C_6mim]Ac and [C_8mim]Ac, respectively.

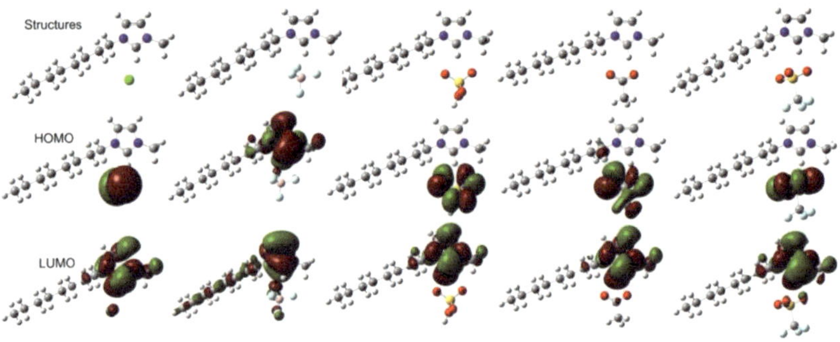

Figure 6.
Equilibrium geometry structures, HOMO and LUMO isosurfaces of [Omim]Y system with B3LYP/6-311++G (d, p) method. From left to right, it is [Omim]Cl, [Omim]BF$_4$, [Omim]HSO$_4$, [Omim]Ac and [Omim]TFO.

ILs	[Omim]Cl	[Omim]BF$_4$	[Omim]HSO$_4$	[Omim]Ac	[Omim]TfO
E_{HOMO}	−5.4007	−8.2613	−5.3992	−5.3951	−5.3943
E_{LUMO}	−0.9120	−1.4030	−0.9009	−0.8979	−0.8985
ΔE	4.4887	6.8583	4.4982	4.4972	4.4958

Table 5.
E_{HOMO}, E_{LUMO} *and* ΔE *(eV) of [Omim]Y with B3LYP/6-311++G(d, p) method.*

and form a feedback bond, which will make the imidazole ring stably adsorb on the iron surface. However, HOMO distribution of [Omim]Cl is very different. The HOMO of [Omim]BF$_4$ is mainly distributed on the imidazole ring and the alkyl group, and only a small amount of anions is distributed. The HOMO of [Omim] HSO$_4$, [Omim]Ac and [Omim]TFO are all allocated on anions, especially O atom. It is further stated that the oxygen-containing group easily forms a coordination bond with the d-orbit of the iron surface and is stably adsorbed on the iron surface.

The E_{HOMO}, E_{LUMO} and the energy gap ΔE of four [Omim]Y ionic liquids are shown in **Table 5**. It is shown in **Table 5** that the order of size of E_{HOMO} and E_{LUMO} is [Omim]TFO > [Omim]Ac > [Omim]HSO$_4$ > [Omim]Cl > [Omim]BF$_4$, which indicates that [Omim]TFO has the strongest ability to give electrons and [Emim]Cl has the strongest ability to get electrons. The order of ΔE is [Omim]BF$_4$ > [Omim] HSO$_4$ > [Omim]Ac > [Omim]TFO > [Omim]Cl, which indicates that [Omim]Cl has the best chemical activity and is the easiest to interact with metal surface.

3.3.2 Global active parameters

Table 6 shows the global activity parameters of [Omim]Y ionic liquids obtained by B3LYP/6-311++G(d, p) method. It is shown in **Table 6** that [Omim]HSO$_4$ has the largest dipole moment (μ), while [Omim]Ac has the smallest dipole moment (μ), indicating that [Omim]HSO$_4$ has the largest molecular polarity and easily interacts with the iron surface. For electronegativity (χ), the maximum of [Omim] BF$_4$ and the second of [Omim]TFO indicate that the ability of [Omim]BF$_4$ and [Omim]TFO to attract electrons is strong, which makes the molecules stably adsorbed on the iron surface by combining with the electrons on the iron surface. Global hardness (η) of [Omim]BF$_4$ is the largest, the global softness (S) is the smallest and the electrophilic index (ω) is the largest, indicating that [Omim]BF$_4$ is a good ionic liquid inhibitor. It can be seen from **Table 6** that the order of polarizability (α) is [Omim]TFO > [Omim]Ac > [Omim]HSO$_4$ > [Omim]BF$_4$ > [Omim]

ILs	μ/Debye	χ/eV	η/eV	S/eV^{-1}	ω	α(a.u.)
[Omim]Cl	12.2083	3.2213	2.0394	0.4903	2.5441	183.8247
[Omim]BF_4	12.2330	4.8321	3.4292	0.2916	3.4045	175.4047
[Omim]HSO_4	12.9717	3.8610	2.5009	0.3999	2.9804	195.8180
[Omim]Ac	9.5918	3.1464	2.2479	0.4449	2.2020	197.2063
[Omim]TfO	12.2888	4.1213	2.6048	0.3839	2.2604	204.3150

Table 6.
Global activity parameters for [Omim]Y system with B3LYP/6-311++G(d, p) method.

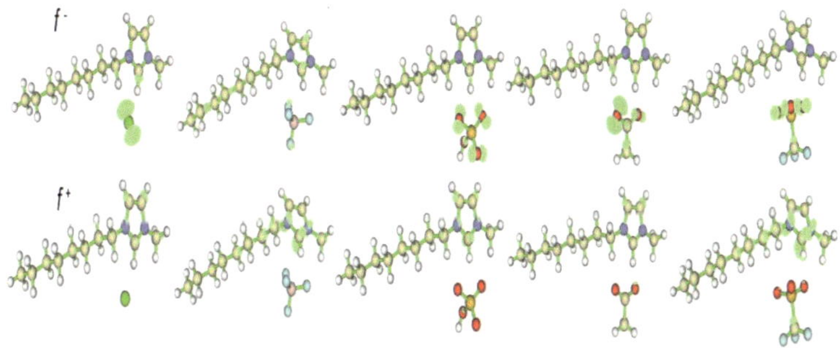

Figure 7.
Fukui index isosurfaces of [Omim]Y ionic liquids with B3LYP/6-311++G(d,p) method. From left to right, it is [Omim]Cl, [Omim]BF_4, [Omim]HSO_4, [Omim]Ac and [Omim]TFO.

Cl, indicating that [Omim]TFO is the most easily adsorbed on the iron surface, with the best corrosion inhibition effect and the highest corrosion inhibition efficiency.

3.3.3 Fukui index distribution

Figure 7 shows the Fukui index distribution of four ionic liquid [Omim]Y systems. From **Figure 7**, it can be seen that the electrophilic attack index of [Omim] Cl is mainly distributed on the Cl atom, and 2C (the number of atoms is shown in **Figure 1**) also has a small amount of the distribution. The nucleophilic attack index is allocated on 1N, 2C, 4C and 5C atoms. The electrophilic attack index of [Omim] BF_4 is 2C, 4C and F, and the electrophilic attack index is mainly distributed on 1N, 2C and 3N. Electric attack indexes of [Omim]HSO_4, [Omim]Ac and [Omim]TFO are mainly distributed on the O atom of anion, while the nucleophilic attack index is mainly distributed on the imidazole ring. In general, we found that the nucleophilic attack site of [Omim]Y is on the imidazole ring, and the electrophilic attack site is mainly on anion, especially O atom.

Compared with the HOMO and LUMO distribution of these three set ionic liquids systems of [$C_n$mim]Cl, [$C_n$mim]Ac and [Omim]Y in **Figures 2**, **4** and **6**, we found that the HOMO distribution is mainly located on the anion containing Cl and O atoms, while the LUMO distribution is mainly located on the imidazole ring of the cation. It shows that the anion containing Cl and O atom, and imidazole ring are the active regions of the ionic liquid. For E_{HOMO} and E_{LUMO}, the E_{HOMO} of [Omim]Cl is the largest, and the E_{LUMO} of [Omim]TFO is the smallest, indicating that [Omim]Cl has the strongest ability to give electrons and [Omim]TFO has the strongest ability to get electrons. The dipole moment (μ) of [Omim]HSO_4 is the largest, which indicates that the molecular polarity is the largest and it is easier to interact with the iron surface. For electronegativity (χ), the maximum is [Omim]BF_4 and the second is [Omim]TFO, which indicates that [Omim]BF_4 and [Omim]TFO have strong ability to attract electrons. The global hardness (η) of [Omim]BF_4 and [Omim]TFO is larger, the global softness (S) is smaller and the electrophilic index (ω) is larger, which indicates that [Omim]BF_4 and [Omim]TFO are ionic liquids with better corrosion inhibition performance. For the polarizability (α), [Omim]TFO is the largest, which indicates that [Omim]TFO is the most easily adsorbed on the iron surface, and its corrosion inhibition effect is the best. In conclusion, [Omim]TFO has the best corrosion inhibition performance and the highest efficiency among the 11 ionic liquids studied.

4. Conclusion

The reaction activities such as E_{LUMO}, E_{HOMO}, the softness (S) and polarizability (α) increase gradually, whereas electronegativity (χ), energy gap (ΔE), hardness (η), dipole moment (μ) and electrophilic index (ω) of three sets ionic liquids 1-alkyl-3-methylimidazole chloride [$C_n$mim]Cl, 1-alkyl-3-methylimidazolium acetate [$C_n$mim]Ac and 1-octyl-3-methylimidazole salt [Omim]Y were studied by density functional theory. The results are as follows:

In [$C_n$mim]Cl and [$C_n$mim]Ac system, increasing the length of the alkyl chain, the HOMO energy E_{HOMO}, LUMO energy E_{LUMO}, softness (S) increase, while the dipole moment (μ), energy gap ΔE, electronegativity (χ), electrophilic index (ω), hardness (η) and polarizability (α) gradually decrease. In [Omim]Y (Y = Cl, BF_4, HSO_4, Ac, TFO) ionic liquids, the structure parameters of ionic liquids vary greatly with different anions.

The active region of [$C_n$mim]Cl system is the imidazole ring and anion Cl^-, while the [$C_n$mim]Ac system is located in the imidazole ring and the O atoms of anion. The active region of [Omim]Y system is located in the imidazole ring and an oxygen-containing group. The reaction is the active region of ionic liquid molecules will help interact with the iron surface and stably adsorb on the iron surface.

For [$C_n$mim]Cl system, the order of inhibition efficiency predicted is [C_8mim]Cl > [C_6mim]Cl > [C_4mim]Cl > [C_2mim]Cl, and [Omim]Cl may have the best inhibition effect. For [$C_n$mim]Ac system, the order of inhibition efficiency is [C_8mim]Ac > [C_6mim]Ac > [C_4mim]Ac > [C_2mim]Ac, and [Omim]Ac may have the best inhibition effect. Inhibition efficiency increases by increasing the alkyl chain length. For [Omim]Y system, [Omim]TFO may have better inhibition efficiency, and the sequence of inhibition efficiency needs further study.

Acknowledgements

This work was supported in part by the National Natural Science Foundation of China (51774158, 5126402) and Back-up Personnel Foundation of Academic and Technology Leaders of Yunnan Province (2011HR013).

Conflict of interest

The authors declare no conflict of interest.

Author details

Guocai Tian* and Weizhong Zhou
State Key Laboratory of Complex Nonferrous Metal Resource Clean Utilization, Faculty of Metallurgical and Energy Engineering, Kunming University of Science and Technology, Yunnan, Kunming, China

*Address all correspondence to: tiangc@kust.edu.cn

References

[1] Zhang TS, Zhang H, Gao H. Inhibitors. Beijing: Chemical Industry Press; 2008. pp. 316-339

[2] Li WH. Synthesis and Evaluation of New Corrosion Inhibitors. Beijing: Science Press; 2015. pp. 1-40

[3] Wen X, Bai P, Luo B, Zheng S, Chen C. Review of recent progress in the study of corrosion products of steels in a hydrogen sulphide environment. Corrosion Science. 2018;**139**:124-140. DOI: 10.1016/j.corsci.2018.05.002

[4] Antonijevic MM, Petrovic MB. Copper corrosion inhibitors. A review. International Journal of Electrochemistry. 2008;**3**:1-28

[5] Li RX. Synthesis and Application of Green Solvent-Ionic Liquids. Beijing: Chemical Engineering Press; 2004. pp. 2-10

[6] Deng Y. Ionic Liquid-Property, Preparation and Application. Beijing: China Petrochemical Press; 2006. pp. 10-23

[7] Zhang SJ, Lv XM, Liu ZP. Ionic Liquids: From Basic Research to Industrial Applications. Beijing: Science Publishing; 2006. pp. 50-380

[8] Verma C, Ebenso EE, Quraishi MA. Ionic liquids as green and sustainable corrosion inhibitors for metals and alloys: An overview. Journal of Molecular Liquids. 2017;**233**:403-414. DOI: 10.1016/j.molliq.2017.02.111

[9] Kesavan D, Gopiraman M, Sulochana N. Green inhibitors for corrosion of metals: A review. Chemical Science Review and Letters. 2012;**1**(1): 1-8

[10] Quan KY. Research on the Method and Application of Strong Polarization Detection of Metal Corrosion. Chongqing: Chongqing University; 2018. pp. 8-12

[11] Verma C, Ebenso EE, Bahadur I, Quraishi MA. An overview on plant extracts as environmental sustainable and green corrosion inhibitors for metals and alloys in aggressive corrosive media. Journal of Molecular Liquids. 2018;**266**:577-590. DOI: 10.1016/j.molliq.2018.06.110

[12] Zarrouk A, Messali M, Zarrok H. Synthesis, characterization and comparative study of new functionalized imidazolium-based ionic liquids derivatives towards corrosion of C38 steel in molar hydrochloric acid. International Journal of Electrochemical Science. 2012;**7**:6998-7015

[13] Zheng X, Zhang S, Li W, Yin L, He J, Wu J. Investigation of 1-butyl-3-methyl-1-H-benzimidazolium iodide as inhibitor for mild steel in sulfuric acid solution. Corrosion Science. 2014;**80**:383-392. DOI: 10.1016/j.corsci.2013.11.053

[14] Kowsari E, Payami M, Amini R, et al. Task-specific ionic liquid as a new green inhibitor of mild steel corrosion. Applied Surface Science. 2014:478-486. DOI: 10.1016/j.apsusc.2013.11.017

[15] Lozano I, Mazario E, Olivaresxometl CO, et al. Corrosion behaviour of API 5LX52 steel in HCl and H_2SO_4 media in the presence of 1,3-dibencilimidazolio acetate and 1,3-dibencilimidazolio dodecanoate ionic liquids as inhibitors. Materials Chemistry and Physics. 2014;**147**(1): 191-197. DOI: 10.1016/j.matchemphys. 2014.04.029

[16] Zheng X, Zhang S, Gong M, et al. Experimental and theoretical study on the corrosion inhibition of mild steel by 1-octyl-3-methylimidazolium l-prolinate in sulfuric acid solution. Industrial & Engineering Chemistry

Research. 2014;**53**(42):16349-16358. DOI: 10.1021/ie502578q

[17] Zhang QB, Hua YX. Corrosion inhibition of mild steel by alkylimidazolium ionic liquids in hydrochloric acid. Electrochimica Acta. 2009;**54**(6):1881-1887. DOI: 10.1016/j.electacta.2008.10.025

[18] Ma Y, Han F, Li Z, et al. Corrosion behavior of metallic materials in acidic-functionalized ionic liquids. ACS Sustainable Chemistry & Engineering. 2016;**4**(2):633-639. DOI: 10.1021/acssuschemeng.5b00974

[19] Olivaresxometl O, Lopezaguilar C, Herrastigonzalez P, et al. Adsorption and corrosion inhibition performance by three new ionic liquids on API 5lA52 steel surface in acid media. Industrial & Engineering Chemistry Research. 2014; **53**(23):9534-9543. DOI: 10.1021/ie4035847

[20] Ma Y, Han F, Li Z, et al. Acidic-functionalized ionic liquid as corrosion inhibitor for 304 stainless steel in aqueous sulfuric acid. ACS Sustainable Chemistry & Engineering. 2016;**4**(9): 5046-5052. DOI: 10.1021/acssuschemeng.6b01492

[21] Feng L. Experimental and theoretical studies of 1-vinyl-3-hexylimidazolium iodide ([VHIM]I) as corrosion inhibitor for the mild steel in sulfuric acid solution. International Journal of Electrochemical Science. 2017:1915-1928. DOI: 10.20964/2017.03.30

[22] Verma C, Obot IB, Bahadur I, et al. Choline based ionic liquids as sustainable corrosion inhibitors on mild steel surface in acidic medium: Gravimetric, electrochemical, surface morphology, DFT and Monte Carlo simulation studies. Applied Surface Science. 2018:134-149. DOI: 10.1016/j.apsusc.2018.06.035

[23] Verma C, Olasunkanmi LO, Bahadur I, et al. Experimental, density functional theory and molecular dynamics supported adsorption behavior of environmental benign imidazolium based ionic liquids on mild steel surface in acidic medium. Journal of Molecular Liquids. 2019;**273**:1-15. DOI: 10.1016/j.molliq.2018.09.139

[24] Arellanes-Lozada P, Olivares-Xometl O, Likhanova NV. Adsorption and performance of ammonium-based ionic liquids as corrosion inhibitors of steel. Journal of Molecular Liquids. 2018;**265**:151-163. DOI: 10.1016/j.molliq.2018.04.153

[25] Likhanova NV, Arellanes-Lozada P, Olivares-Xometl O. Effect of organic anions on ionic liquids as corrosion inhibitors of steel in sulfuric acid solution. Journal of Molecular Liquids. 2019;**279**:267-278. DOI: 10.1016/j.molliq.2019.01.126

[26] Gece G. The use of quantum chemical methods in corrosion inhibitor studies. Corrosion Science. 2008;**50**(11): 2981-2992. DOI: 10.1016/j.corsci.2008.08.043

[27] Murulana LC, Singh AK, Shukla SK, Kabanda MM, Ebenso EE. Experimental and quantum chemical studies of some bis(trifluoromethyl-sulfonyl) imide imidazolium- based ionic liquids as corrosion inhibitors for mild steel in hydrochloric acid solution. Industrial & Engineering Chemistry Research. 2012; **51**(40):13282-13299. DOI: 10.1021/ie300977d

[28] Sasikumar Y, Adekunle AS, Olasunkanmi LO, Bahadur I, Baskar R, Kabanda MM, et al. Experimental, quantum chemical and Monte Carlo simulation studies on the corrosion inhibition of some alkyl imidazolium ionic liquids containing tetrafluoroborate anion on mild steel in acidic medium. Journal of Molecular Liquids. 2015;**211**:105-118. DOI: 10.1016/j.molliq.2015.06.052

[29] Mashuga ME, Olasunkanmi L, Adekunle AS, Yesudass S, Kabanda MM, Ebenso EE. Adsorption, thermodynamic and quantum chemical studies of 1-hexyl-3-methyl-imidazolium based ionic liquids as corrosion inhibitors for mild steel in HCl. Materials. 2015;**8**(6): 3607-3632. DOI: 10.3390/ma8063607

[30] Yesudass S, Olasunkanmi LO, Bahadur I, Kabande MM, Obot IB, Ebenso EE. Experimental and theoretical studies on some selected ionic liquids with different cations/anions as corrosion inhibitors for mild steel in acidic medium. Journal of the Taiwan Institute of Chemical Engineers. 2016;**64**:252-268. DOI: 10.1016/j.jtice.2016.04.006

[31] Frisch MJ, Trucks GW, Schlegel HB, Scuseria GE, Robb MA, Cheeseman JR, et al. Gaussian 09, Revision B.02. Wallingford CT: Gaussian, Inc.; 2009

[32] Lu T, Chen F. Multiwfn: A multifunctional wavefunction analyzer. Journal of Computational Chemistry. 2012;**33**(5):580-592. DOI: 10.1002/jcc.22885

[33] Lu T, Chen F. Quantitative analysis of molecular surface based on improved marching tetrahedra algorithm. Journal of Molecular Graphics and Modelling. 2012;**38**:314-323. DOI: 10.1016/j.jmgm.2012.07.004

[34] Zhao HX, Zhang XH, Ji L, Hu HX, Li QS. Quantitative structure–activity relationship model for amino acids as corrosion inhibitors based on the support vector machine and molecular design. Corrosion Science. 2014;**83**: 261-271. DOI: 10.1016/j.corsci.2014.02.023

[35] Daoud D, Douadi T, Hamani H, Chafaa S, Al-Noaimi M. Corrosion inhibition of mild steel by two new S-heterocyclic compounds in 1 M HCl: Experimental and computational study. Corrosion Science. 2015;**94**:21-37. DOI: 10.1016/j.corsci.2015.01.025

[36] Koopmans T. Ordering of wave functions and eigenenergies to the individual electrons of an atom. Physica. 1933;**1**(1):104-113

[37] Yang W, Parr RG. Hardness, softness, and the Fukui function in the electronic theory of metals and catalysis. Proceedings of the National Academy of Sciences of the United States of America. 1985;**82**(20):6723-6726. DOI: 10.1073/pnas.82.20.6723

[38] Chermette H. Chemical reactivity indexes in density functional theory. Journal of Computational Chemistry. 1999;**20**(1):129-154. DOI: 10.1002/(sici)1096-987x(19990115)20:1<129::aid-jcc13>3.0.co;2-a

[39] Obot IB, Macdonald DD, Gasem ZM. Density functional theory (DFT) as a powerful tool for designing new organic corrosion inhibitors. Part 1: An overview. Corrosion Science. 2015; **99**:1-30. DOI: 10.1016/j.corsci.2015.01.037

[40] Li XH, Deng SD, Fu H, Li TH. Adsorption and inhibition effect of 6-benzyl-aminopurine on cold rolled steel in 1.0 M HCl. Electrochimica Acta. 2009;**54**(16):4089-4098. DOI: 10.1016/j.electacta.2009.02.084

[41] Parr RG, Chattaraj PK. Principle of maximum hardness. Journal of the American Chemical Society. 1991; **113**(5):1854-1855. DOI: 10.1021/ja00005a072

[42] Parr R, Yang W. Density Functional Theory of Atoms and Molecules. New York: Oxford Science Publications; 1989

[43] Guo L, Ren XL, Zhou Y, Xu SY, Gong YL, Zhang ST. Theoretical evaluation of the corrosion inhibition performance of 1,3-thiazole and its amino derivatives. Arabian Journal of Chemistry. 2017;**10**(1):121-130. DOI: 10.1016/j.arabjc.2015.01.005

[44] Shahraki M, Dehdab M, Elmi S. Theoretical studies on the corrosion inhibition performance of three amine derivatives on carbon steel: Molecular dynamics simulation and density functional theory approaches. Journal of the Taiwan Institute of Chemical Engineers. 2016;**62**:313-321. DOI: 10.1016/j.jtice.2016.02.010

[45] Fukui K. Role of frontier orbitals in chemical reactions. Science. 1982; **218**(4574):747-754. DOI: 10.1126/science.218.4574.747

[46] Saha SK, Hens A, Murmu NC, Banerjee P. A comparative density functional theory and molecular dynamics simulation studies of the corrosion inhibitory action of two novel N-heterocyclic organic compounds along with a few others over steel surface. Journal of Molecular Liquids. 2016;**215**:486-495. DOI: 10.1016/j.molliq.2016.01.024

Section 4

Chemoinformatic

Chapter 4

Conceptual DFT as a Helpful Chemoinformatics Tool for the Study of the Clavanin Family of Antimicrobial Marine Peptides

Norma Flores-Holguín, Juan Frau and Daniel Glossman-Mitnik

Abstract

A well-behaved model chemistry previously validated for the study of the chemical reactivity of peptides was considered for the calculation of the molecular properties and structures of the clavanin family of antimicrobial marine peptides. A methodology based on conceptual density functional theory (CDFT) was chosen for the determination of the reactivity descriptors. The molecular active sites were associated with the active regions of the molecules related to the nucleophilic and electrophilic Fukui functions. Finally, the drug-likenesses and the bioactivity scores for the clavanin peptides were predicted through a homology methodology relating them with the calculated reactivity descriptors, while other properties like the pKas were determined following a methodology developed by our group.

Keywords: clavanin, conceptual DFT, chemical reactivity, drug-likeness features, bioactivity scores

1. Introduction

Bioactive peptides are promising novel drug leads that may fill the gap between small molecules and larger biologicals. This is reflected by a multitude of recent peptide discovery and development approaches. However, their use as therapeutic lead molecules is challenged by their typically poor stability and lack of oral bioavailability. This is often due to the linear nature of peptides that not only exhibit free ends but multiple cleavage sites that are readily recognized by enzymes that degrade peptide chains into inactive fragments or single amino acids [1–10].

The marine environment is considered Earth's last frontier of exploration. In fact, a common belief is that just less than 5% of the vast and rich marine environment has been explored. Our seas and oceans represent a very unknown resource for the discovery of novel organisms, (bio)products, (bio)processes, and for the development of bioinspired synthetic drugs. Recent advances in genetics and other (bio)molecular techniques are providing all necessary tools to access these still-untapped marine resources on a larger scale and, consequently, enable exploitation of the true promise of the blue biotechnology [11].

Clavanins are α-helical antimicrobial peptides isolated from a mixed tunicate population of *Styela clava* hemocytes. Clavanin A, VFQFLGKIIH10HVGNFVHGFS 20HVFa, and the clavanins BE are 23-peptide antibiotics containing 18 identical residues bearing a C-terminally amidated amino acid moiety [12].

This study seeks to assess the molecular properties such as the chemical reactivity and bioactivity scores through the application of the conceptual density functional theory (CDFT) concepts. Understanding the reactivity properties of the clavanin family of antimicrobial peptides is important in the use of Fukui functions to represent the peptide reactivity with the molecular systems in the process of developing new drugs. Bioavailability and bioactivity scores of the molecular systems will also be compared with the descriptors of the conceptual density functional theory [13–16].

2. Computational methodology

The generation of 3D structures and the proposition of their respective low-energy conformers in the prediction and calculation of the properties of the five members of the clavanin family of antimicrobial peptides in this study was carried out using ChemAxon Calculator Plugins. In the process of geometry optimization, the molecules with the lowest energy conformation were used, while the DFTBA program was used in the optimization of the rest of the conformers. The MN12SX/Def2TZVP/H20 model chemistry was used in the re-optimization of the five conformers having the lowest energy. Consequently, the real minimum approach was used in the confirmation of the optimized clavanin structures through the application of the vibrational frequency analysis technique. The process of calculating the electronic properties for the chemical reactivity of the antimicrobial peptides involved the use of MN12SX/Def2TZVP/H2O model chemistry through the optimized molecular structures.

3. Results and discussion

ChemAxon Calculator Plugins were used in the process of deriving the molecular structures and the bioactivity properties of the conformers. The optimization and re-optimization of the conformers was carried out using the DFTBA program and the MN12SX/DefTZVP/H20 model chemistry, respectively, as explained in the Computational Methodology section [17]. The graphical sketches of the molecular structures of the clavanins A–E are shown in **Figure 1**. The density functional tight-binding method was used in the re-optimization of the molecular structures, while the MN12SX density functional method combined with the SMD solvent model and the DefTZVP were used in the second optimization of the molecular structures. The MN12SX/DefTZVP/H20 model chemistry was used in determining the electronic properties of each molecular structure after using calculation analysis procedures to determine whether all the molecular structures correspond to their respective minimum energy requirements. According to Becke, a common misconception exists in the connection of the electronic ground states and the adiabatic connection since the superiority of the Kohn-Sham (or KS) model is not recognized due to its minimum molecular energy [18]. Baerends et al. state that the level of energy excitation within a KS system is used as an effective measure of the optimization to the molecular optical gap [19]. Thus, the HOMO-LUMO gap of the KS model is used to approximate the excitation energy within the KS model based on which is a basic requirement in determining the consistency with the molecular structures [20].

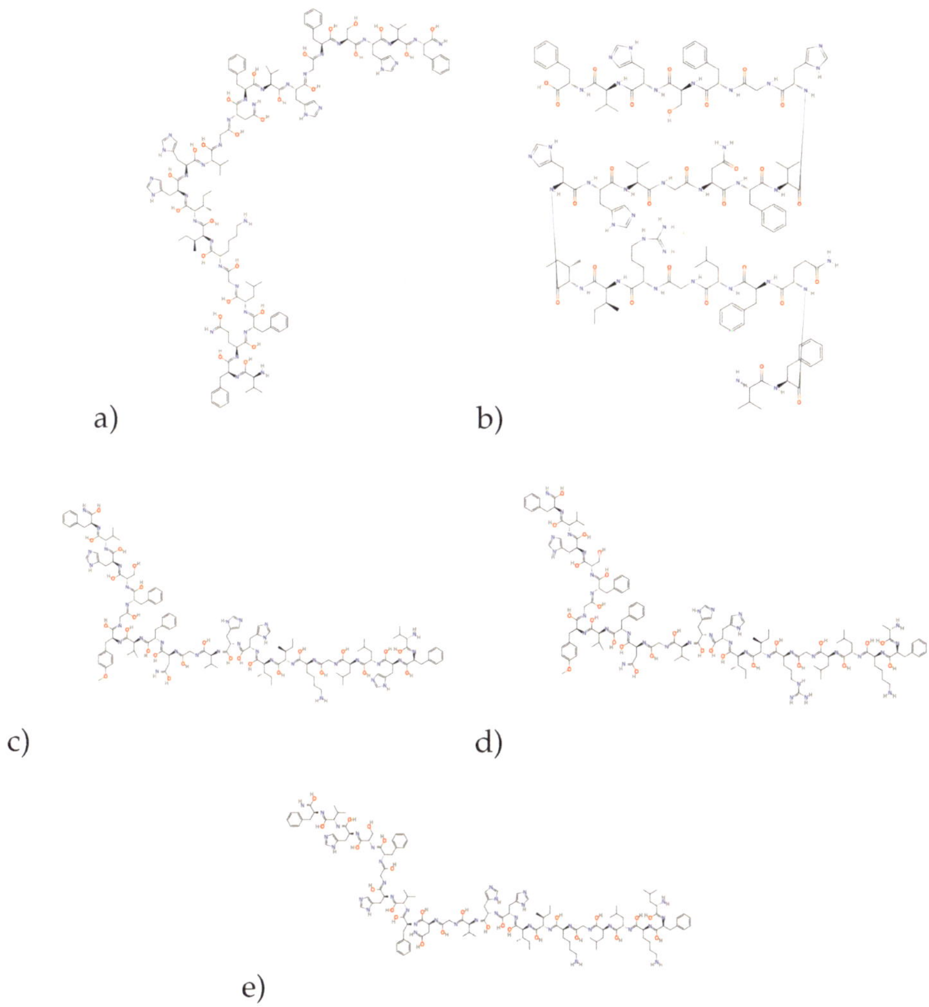

Figure 1.
Graphical representation of the optimized molecular structures of the marine antimicrobial peptides of the clavanin family: (a) Clavanin A, (b) Clavanin B, (c) Clavanin C, (d) Clavanin D, and (e) Clavanin E.

Ground state calculations are used in determining the optimal maximum absorption wavelength that belongs to the marine peptides of the clavanin family based on the predetermined density functional to find the respective λ_{max} values through the application of theoretical models to establish the HOMO-LUMO gaps.

Therefore, the calculation of the maximum wavelength absorption of the clavanins A–E antimicrobial peptides involved conducting ground-state calculations with the aforementioned density functional at the same level of model chemistry and theory and determining the HOMO-LUMO gap.

The amino acid sequences for the Clavanins A–E are shown as follows [21]:

	10	**20**	
Clavanin A	VFQFLGKIIH	HVGNFVHGFS	HVF
Clavanin B	VFQFLGRIIH	HVGNFVHGFS	HVF
Clavanin C	VFHLLGKIIH	HVGNFVYGFS	HVF
Clavanin D	AFKLLGRIIH	HVGNFVYGFS	HVF
Clavanin E	LFKLLGKIIH	HVGNFVHGFS	HVF

The calculated electronic properties, namely, the total electronic energy, the energies of the HOMO and LUMO orbitals, and the maximum absorption wavelength (in nm) calculated on the basis of the HOMO-LUMO gap of the ground state and determined by using the MN12SX/Def2TZVP/H2O model chemistry are presented in **Table 1**.

3.1 Calculation of global reactivity descriptors

According to Frau and Glossman-Mitnik, the evaluation of marine peptides and melanoidins in the generation of HOMO and LUMO energies is required in the verification of the levels of agreement with the estimated Koopmans' theorem based on the combination of the MN12SX density functional and the Def2TZVP basis set [22–28]. This justifies the application of the proposed KID (for Koopmans in DFT) technique.

Considering the KID technique used on the previous studies being integrated into the finite difference approximation [22–28], the following expressions can be used to define the global reactivity descriptors [15, 16, 29, 30]:

$$\text{Electronegativity}\ \chi = -\frac{1}{2}(I + A) \approx \frac{1}{2}(\varepsilon_L + \varepsilon_H) \quad (1)$$

$$\text{Global hardness}\ \eta = (I - A) \approx (\varepsilon_L - \varepsilon_H) \quad (2)$$

$$\text{Electrophilicity}\ \omega = \frac{\mu^2}{2\eta} = \frac{(I + A)^2}{4(I - A)} \approx \frac{(\varepsilon_L + \varepsilon_H)^2}{4(\varepsilon_L - \varepsilon_H)} \quad (3)$$

$$\text{Electrodonating power}\ \omega^- = \frac{(3I + A)^2}{16(I - A)} \approx \frac{(3\varepsilon_H + \varepsilon_L)^2}{16\eta} \quad (4)$$

$$\text{Electroaccepting power}\ \omega^+ = \frac{(I + 3A)^2}{16(I - A)} \approx \frac{(\varepsilon_H + 3\varepsilon_L)^2}{16\eta} \quad (5)$$

$$\text{Net electrophilicity}\ \Delta\omega^{\pm} = \omega^+ - (-\omega^-) = \omega^+ + \omega^- \quad (6)$$

where ε_H and ε_L are the HOMO and LUMO energies associated with each of the peptides.

The calculated values for these global reactivity descriptors using the MN12SX/Def2TZVP/H2O model chemistry and the associated HOMO and LUMO energies are displayed in **Table 2**.

	Total electronic energy	HOMO	LUMO	λ_{max}
Clavanin A	−8960.8018	−6.1098	−0.6329	226
Clavanin B	−9090.5934	−6.2148	−1.2218	248
Clavanin C	−8985.1955	−6.1046	−0.6697	228
Clavanin D	−8964.4080	−6.1623	−0.6776	226
Clavanin E	−8852.4122	−6.1231	−0.6555	227

Table 1.
Electronic energies of the neutral molecular systems (in au) of the clavanins A–E, the HOMO and LUMO orbital energies (in eV), and the maximum absorption wavelengths λ_{max} (nm) calculated with the MN12SX density functionals and the Def2TZVP basis set using water as solvent simulated with the SMD parametrization of the IEF-PCM model.

3.2 Calculation of the pKa of the clavanins A–E peptides

The previous discussion focused on the application of the conceptual DFT descriptors to evaluate the computation prediction of the pKas peptides where it was established that the pKa = 16.3088–0.868 η relationship would play an important role in the initial prediction of complex peptides, which are important in the manufacture of medical drugs [31]. Given the biological level of pH, the peptides under study exist as neutral molecules and are still considered to be neutral during the pKa computations [31]. The pKa relationship is also important in the optimization of the molecular structure of every conformer as well as the computation of the pKa values for all molecules given the η values shown in **Table 2**. The computational results of the pKa values for the clavanin molecules are shown in **Table 3**.

The pKa values shown in **Table 3** indicate that the computational methodology used is effective in the differentiation of the respective pKa values for all the peptidic molecules irrespective of the significance of the difference. The pKa values of these peptides are important in the manufacture of pharmaceutical drugs by explaining the procedures used in drug delivery and their respective action mechanisms.

Molecule	Electronegativity	Global hardness	Electrophilicity
Clavanin A	3.3714	5.4768	1.0376
Clavanin B	3.7183	4.9930	1.3845
Clavanin C	3.3871	5.4349	1.0555
Clavanin D	3.4199	5.4847	1.0662
Clavanin E	3.3893	5.4676	1.0505
Molecule	**Electrodonating power**	**Electroaccepting power**	**Net electrophilicity**
Clavanin A	4.1033	0.7319	4.8352
Clavanin B	4.9402	1.2219	6.1622
Clavanin C	4.1442	0.7570	4.9012
Clavanin D	4.1852	0.7653	4.9505
Clavanin E	4.1374	0.7481	4.8855

Table 2.
Global reactivity descriptors of the clavanins A–E molecules calculated with the MN12SX/Def2TZVP/H2O model chemistry.

Molecule	pKa
Clavanin A	11.78
Clavanin B	12.18
Clavanin C	11.82
Clavanin D	11.77
Clavanin E	11.79

Table 3.
pKas of the clavanins A–E molecules.

3.3 Local reactivity descriptor calculation

Applying the same ideas as before, the definitions for the local reactivity descriptors will be [15]:

$$\text{Nucleophilic Fukui function} f^{+}(\mathbf{r}) = \rho_{N+1}(\mathbf{r}) - \rho_{N}(\mathbf{r}) \quad (7)$$

$$\text{Electrophilic Fukui function} f^{-}(\mathbf{r}) = \rho_{N}(\mathbf{r}) - \rho_{N-1}(\mathbf{r}) \quad (8)$$

which are relationships between the electronic densities of the neutral, positive, and negative species.

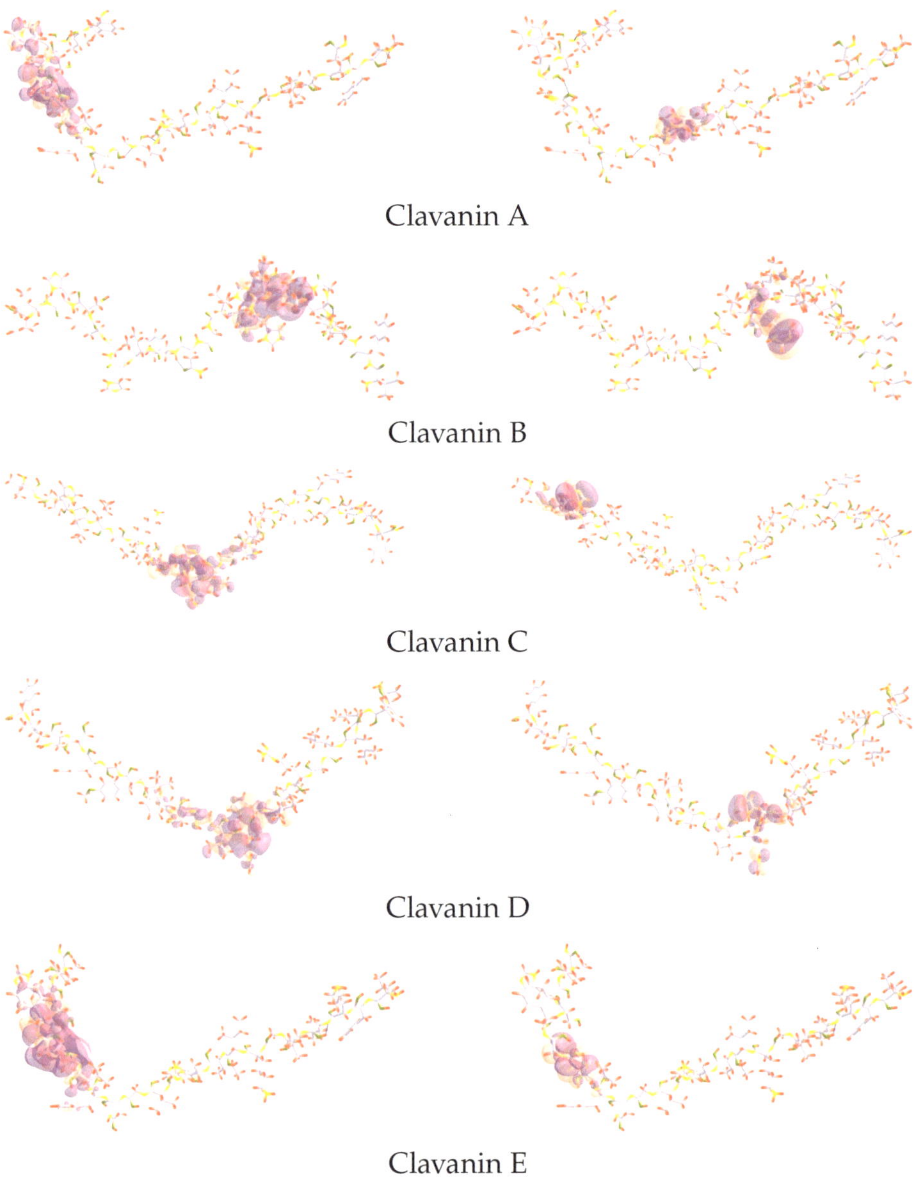

Figure 2.
The right column represents the nucleophilic Fukui function $f^{+}(\mathbf{r})$, while the left column represents the electrophilic Fukui function $f^{-}(\mathbf{r})$ of five antimicrobial peptides of marine origin belonging to the clavanin family.

The electrophilic Fukui functions $f^-(\mathbf{r})$ and nucleophilic Fukui functions $f^+(\mathbf{r})$ for the clavanin A–E molecules are shown in **Figure 2**.

3.4 Bioavailability and bioactivity scores

According to Leeson et al., it is important to check the species level of compliance of a potential therapeutic drug to the Lipinski rule of five, which explains whether the compound contains certain drug properties [32]. Molinspiration and MolSoft software were used to compute the molecular drug properties in a compound as shown in **Table 4** given that miLogP is a representation of the water partition coefficient. The rate of violations of the Lipinski rule of five is measured using nviol, while TPSA represents the polar surface area of the molecule. The hydrogen bond donors and the hydrogen bond acceptors are represented by nOHNH and nON, respectively.

Table 4 shows the application of the Lipinski rule of five in the calculation of the molecular properties of the antimicrobial marine peptides of the clavanin family. Volume represents the molecular volume of the peptides, while MW represents the molecular weight of the peptides.

The degree of oral bioavailability of the marine molecules that can be potentially used in the manufacture of drugs is measured using the Lipinski rule of five by determining the molecules that possess drug-like properties. However, this technique cannot be applied in measuring the bioavailability of the peptides due to the existence of hydrogen bonds and molecular weight properties as shown in **Table 4**.

This study applied a different technique in the evaluation of the chemical structure of other compounds that were predicted to possess similar pharmacological properties as the clavanin peptides under study. As illustrated in the Computational Methodology section, the evaluation of the pharmacological properties of different compounds in the process of determining the bioactivity scores can be carried out using Molinspiration software based on the variability of the drug targets as shown in **Table 5**. According to the table, the organic molecules whose bioactivity score is less than zero are considered to be active, while the organic molecules whose bioactivity score are between zero and negative five are considered to be moderately active, and the organic molecules with a score of less than negative five are considered to be inactive. All peptides that were studied during this study were found to have moderate bioactivity scores with respect to enzymatic reactivity.

Property	Clavanin A	Clavanin B	Clavanin C	Clavanin D	Clavanin E
miLogP	−0.95	−4.49	−1.18	−0.23	−0.30
TPSA	1036.27	986.53	1001.42	1034.64	1018.21
natoms	192	194	193	192	190
nON	61	63	60	61	60
nOHNH	38	37	35	38	37
Nviol	3	3	3	3	3
Nrotb	83	84	83	85	85
Volume	2460.93	2477.80	2489.31	2483.28	2470.86
MW	2666.14	2695.14	2681.20	2672.19	2646.19

Table 4.
Molecular properties of the clavanins A–E peptides calculated to verify the Lipinski rule of five.

Property	Clavanin A	Clavanin B	Clavanin C	Clavanin D	Clavanin E
GPCR ligand	−4.18	−4.17	−4.18	−4.17	−4.17
Ion channel Modulator	−4.19	−4.18	−4.19	−4.18	−4.18
Kinase inhibitor	−4.20	−4.20	−4.20	−4.20	−4.19
Nuclear receptor ligand	−4.21	−4.21	−4.21	−4.21	−4.21
Protease inhibitor	−4.17	−4.15	−4.18	−4.16	−4.17
Enzyme inhibitor	−4.18	−4.17	−4.17	−4.20	−4.17

Table 5.
Bioactivity scores of the clavanins A–E peptides.

Table 5 shows the bioactivity scores of the clavanin family of antimicrobial marine peptides based on the interactions with various enzyme inhibitors, GPCR ligand, the protease inhibitors, the ion channel modulators, the kinase inhibitors, and the nuclear receptors.

4. Conclusions

Along this research, the chemical reactivity of a group of five members of the clavanin family of marine antimicrobial peptides was studied by resorting to the conceptual DFT as a tool to explain the molecular interactions.

The information about the global and local reactivity descriptors of the antimicrobial peptides acquired in this work could be helpful to assist in the design of new pharmaceutical drugs based on these compounds.

Among the many descriptors that could be useful for the development of new medicines, the pKa is of paramount importance because it is related to the water solubility of drugs. Thus, when the experimental values of the pKa are unknown, the approximate QSAR relationship employed in this work could be a useful predictive tool for the determination of the pKas of small and large peptides.

Finally, the molecular properties related to bioavailability have been predicted using different methodologies already described in the literature, and the descriptors used for the quantification of the bioactivity allowed the characterization of the studied peptides.

Acknowledgements

Norma Flores-Holguín and Daniel Glossman-Mitnik are researchers of CIMAV and CONACYT. Daniel Glossman-Mitnik conducted this work while being a visiting lecturer at the University of the Balearic Islands from which support is gratefully acknowledged.

Conflict of interest

The authors declare no conflict of interest.

Author details

Norma Flores-Holguín[1†], Juan Frau[2†] and Daniel Glossman-Mitnik[1*†]

1 Laboratorio Virtual NANOCOSMOS, Departamento de Medio Ambiente y Energía, Centro de Investigación en Materiales Avanzados, Chihuahua, Mexico

2 Departament de Química, Universitat de les Illes Balears, Palma de Mallorca, Spain

*Address all correspondence to: daniel.glossman@cimav.edu.mx

† These authors contributed equally.

References

[1] Koehbach J. Structure-activity relationships of insect defensins. Frontiers in Chemistry. 2017;**5**:45

[2] O'Brien-Simpson NM, Hoffmann R, Chia CSB, Wade JD, editors. Antimicrobial and Anticancer Peptides. Switzerland: Frontiers Media SA; 2018

[3] Wang G. Antimicrobial Peptides: Discovery, Design and Novel Therapeutic Strategies. Wallingford, Oxfordshire, UK/Cambridge, MA: CABI; 2010

[4] Reen F, Gutiérrez-Barranquero J, Dobson A, Adams C, O'Gara F. Emerging concepts promising new horizons for marine biodiscovery and synthetic biology. Marine Drugs. 2015; **13**(5):2924-2954

[5] Gopal R, Lee J, Kim Y, Kim M-S, Seo C, Park Y. Anti-microbial, anti-biofilm activities and cell selectivity of the NRC-16 peptide derived from witch flounder, *Glyptocephalus cynoglossus*. Marine Drugs. 2013;**11**(6):1836-1852

[6] Falanga A, Lombardi L, Franci G, Vitiello M, Iovene M, Morelli G, et al. Marine antimicrobial peptides: Nature provides templates for the design of novel compounds against pathogenic bacteria. International Journal of Molecular Sciences. 2016;**17**(5):785

[7] Lee IH, Zhao C, Cho Y, Harwig SS, Cooper EL, Lehrer RI. Clavanins, α-helical antimicrobial peptides from tunicate hemocytes. FEBS Letters. 1997; **400**(2):158-162

[8] In HL, Cho Y, Lehrer RI. Effects of pH and salinity on the antimicrobial properties of Clavanins. Infection and Immunity. 1997;**65**(7):2898-2903

[9] van Kan EJM, van der Bent A, Demel RA, de Kruijff B. Membrane activity of the peptide antibiotic Clavanin and the importance of its glycine residues. Biochemistry. 2001; **40**(21):6398-6405

[10] van Kan EJM, Ganchev DN, Snel MME, Chupin V, van der Bent A, de Kruijff B. The peptide antibiotic Clavanin a interacts strongly and specifically with lipid bilayers. Biochemistry. 2003;**42**(38):11366-11372

[11] Rampelotto P. Grand Challenges in Marine Biotechnology. Cham: Springer; 2018

[12] Jakubke H-D, Sewald N. Peptides from A to Z: A Concise Encyclopedia. Weinheim: Wiley-VCH Verlag GmbH & Co.KGaA; 2008

[13] Gupta GK, Kumar V. Chemical Drug Design. Berlin: Walter de Gruyter GmbH; 2016

[14] Gore M, Jagtap UB. Computational Drug Discovery and Design. New York: Springer Science+Business Media, LLC; 2018

[15] Parr RG, Yang W. Density-Functional Theory of Atoms and Molecules. New York: Oxford University Press; 1989

[16] Geerlings P, De Proft F, Langenaeker W. Conceptual density functional theory. Chemical Reviews. 2003;**103**:1793-1873

[17] Frisch MJ, Trucks GW, Schlegel HB, Scuseria GE, Robb MA, Cheeseman JR, et al. Gaussian 09 Revision E.01. Wallingford, CT: Gaussian Inc; 2016

[18] Becke AD. Vertical excitation energies from the adiabatic connection. The Journal of Chemical Physics. 2016; **145**(19):194107

[19] Baerends EJ, Gritsenko OV, van Meer R. The Kohn-Sham gap, the

fundamental gap and the optical gap: The physical meaning of occupied and virtual Kohn-Sham orbital energies. Physical Chemistry Chemical Physics. 2013;**15**(39):16408-16425

[20] van Meer R, Gritsenko OV, Baerends EJ. Physical meaning of virtual Kohn-Sham orbitals and orbital energies: An ideal basis for the description of molecular excitations. Journal of Chemical Theory and Computation. 2014;**10**(10):4432-4441

[21] Rajanbabu V, Chen J-Y, Jen-Leuh W. Chapter 31: Antimicrobial peptides from marine organisms. In: Kim S-K, editor. Springer Handbook of Marine Biotechnology. Berlin Heidelberg: Springer-Verlag; 2015. pp. 747-758

[22] Frau J, Glossman-Mitnik D. Molecular reactivity and absorption properties of melanoidin blue-G1 through conceptual DFT. Molecules. 2018;**23**(3):559-515

[23] Frau J, Glossman-Mitnik D. Conceptual DFT study of the local chemical reactivity of the dilysyldipyrrolones A and B intermediate melanoidins. Theoretical Chemistry Accounts. 2018;**137**(5):1210

[24] Frau J, Glossman-Mitnik D. Conceptual DFT study of the local chemical reactivity of the colored BISARG melanoidin and its protonated derivative. Frontiers in Chemistry. 2018;**6**(136):1-9

[25] Frau J, Glossman-Mitnik D. Molecular reactivity of some Maillard reaction products studied through conceptual DFT. Contemporary Chemistry. 2018;**1**(1):1-14

[26] Frau J, Glossman-Mitnik D. Computational study of the chemical reactivity of the blue-M1 intermediate melanoidin. Computational and Theoretical Chemistry. 2018;**1134**:22-29

[27] Frau J, Glossman-Mitnik D. Chemical reactivity theory applied to the calculation of the local reactivity descriptors of a colored Maillard reaction product. Chemical Science International Journal. 2018;**22**(4):1-14

[28] Frau J, Glossman-Mitnik D. Blue M2: An intermediate melanoidin studied via conceptual DFT. Journal of Molecular Modeling. 2018;**24**(138):1-13

[29] Gázquez JL, Cedillo A, Vela A. Electrodonating and electroaccepting powers. Journal of Physical Chemistry A. 2007;**111**(10):1966-1970

[30] Chattaraj PK, Chakraborty A, Giri S. Net electrophilicity. Journal of Physical Chemistry A. 2009;**113**(37): 10068-10074

[31] Frau J, Hernández-Haro N, Glossman-Mitnik D. Computational prediction of the pKas of small peptides through conceptual DFT descriptors. Chemical Physics Letters. 2017;**671**: 138-141

[32] Leeson P. Drug discovery: Chemical beauty contest. Nature. 2012;**481**(7382): 455-456

Section 5

Molecular Docking

Chapter 5

DFT and Molecular Docking Studies of a Set of Non-Steroidal Anti-Inflammatory Drugs: Propionic Acid Derivatives

Safna Hussan Kodakkat Parambil,
Hisana Asharaf Thozhuvana Parambil,
Shahina Parammal Hamza,
Anjali Thirumangalath Parameswaran,
Mohamed Shahin Thayyil and Muraleedharan Karuvanthodi

Abstract

Inflammation is the body's defense mechanism to eradicate the spread of injurious agents in the affected mammalian tissues with a number of cellular mediators. Nonsteroidal anti-inflammatory drugs (NSAIDs) are the most commonly used drugs worldwide in such situations. The mode of action of the non-steroid anti-inflammatory drugs (NSAIDs) is attributed primarily to the inhibition of prostaglandin (PG) synthesis, and more specifically, to the inhibition of the COX enzyme system. This work can be considered as an effort to gain a deeper insight into the physiochemical properties of a few well-known NSAIDs namely; ketoprofen, fenoprofen, flurbiprofen and ibuprofen. A quantum computational approach was used to predict geometry, molecular electrostatic potential (MESP), polarizability, hyperpolarizability and molecular docking study of all selected NSAIDs with human COX-1 and COX-2 enzymes were done to predict the most active drug among the four and to demonstrate good selectivity profile with COX enzymes.

Keywords: organic chemistry, theoretical chemistry, pharmaceutical chemistry, DFT, molecular docking, propionic acid derivatives

1. Introduction

Nonsteroidal anti-inflammatory drugs (NSAID) are a class of drugs that reduce pain, decrease fever, prevent blood clots and, in higher doses decrease inflammations too. NSAIDs have been widely used to treat a number of diseases such as heart disease, various cancers, and Alzheimer's, pathogenic conditions. The term nonsteroidal distinguishes these drugs from steroids, which having a similar eicosanoid-depressing, anti-inflammatory action and have a broad range of other effects [1]. NSAIDs obstruct the generation of prostaglandins (chemical messengers that regulate inflammation, fever, and the sensation of pain) by restraining the activity of a

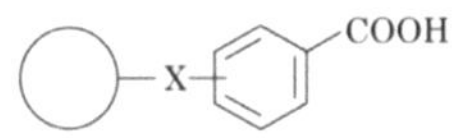

Figure 1.
Chemical structure of propionic acid derivatives.

compound, cyclooxygenase (COX); COX-1 and COX-2. Both the COX-1 and COX-2 enzymes serve important homeostatic roles in the human body. Depending on their chemical structures, NSAIDs are broadly divided into two major classes like non-selective COX inhibitors and selective COX-2 inhibitors. The classification based on the chemical structure is non-selective COX inhibitors and selective COX-2 inhibitors [2–5]. The non-selective COX inhibitors are salicylates, propionic acid derivatives, enolic acid (oxicam) derivatives, anthranilic acid derivatives, selective COX-2 inhibitors, sulfonanilide, and others. In which, COX-1 is considered as important for the production of prostaglandins of homeostatic maintenance, such as platelet aggregation, the regulation of blood flow in the kidney and stomach, and the regulation of gastric acid secretion. While COX-2 is considered as an inducible isoenzyme, although there is some constitutive expression in the kidney, brain, bone, female reproductive system, and gastrointestinal (GI) tract. Thus, the COX-2 is an enzyme plays an important role in pain and inflammatory processes [1–3, 6].

The profen drugs are a category of nonselective, non-steroidal anti-inflammatory drugs (NSAIDs), which reduce pain (analgesia), body temperature during fever (antipyretic), signs of inflammation (anti-inflammatory activity), and in mice, slow the development of cancers. They are one of the most commonly prescribed pain medications. The profen drugs are derivatives of 2-phenylpropanoic acid. In this work, we have preferred such propionic acid derivatives. 2-phenylpropanoic acid- profen drugs and their general chemical structure are depicted in **Figure 1**. Few drugs under this category are; ibuprofen, ketoprofen, naproxen, fenoprofen, flurbiprofen, and oxaprozin. In an effort to elucidate a more deeper insight on the physicochemical properties of 2-phenylpropanoic acid- profen drugs we present a detailed discussion on quantum computational calculations and predictions based on their structural geometry, frontier molecular orbitals non-linear optical properties (NLO), of all selected compounds, were done using B3LYP/6311G++(d,p) level of theory. In addition, the computationally calculated electronic properties such as Highest Occupied Molecular Orbital (HOMO) and Lowest Unoccupied Molecular Orbital (LUMO), Bond Dissociation Enthalpy (BDE), ionization potential (IP), electron affinity (EA), hardness (η), softness (S), electronegativity (χ) and electrophilic index (ω) were also calculated to get an insight into its property by means of its anti-inflammatory activities. This study will offer knowledge of their action and also help us to design new drugs with therapeutic effects, experimental and thus computational studies are of interest for the rationale of the action mechanism of bioactive compounds. Further, in order to have a better understanding about the interaction with target proteins, molecular docking was also conducted by determining the probable binding modes of it by inserting all selected ligands into the active sites of the COX enzymes.

2. Materials and method

2.1 Materials

The input structures the drugs; ketoprofen (PubChem: 3825), ibuprofen (PubChem: 3672), fenoprofen (PubChem: 3342) and flurbiprofen (PubChem: 3394) were taken from the PubChem database [7] which are in SDF (Standard Data File) format and were converted to GJF (Gaussian Job File) input files using the application Open Babel [8].

2.2 Computational details

All the quantum calculations have been performed by density functional theory using a Gaussian 09 software package [9]. The initial geometries chosen for calculation was taken from the PubChem database and optimized with B3LYP/6311G++(d,p) level of the theory [7]. The B3LYP is Becke's three-parameter practical hybrid methods that add the exchange and electronic correlation terms in DFT, including the Lee, Yang Parr (LYP) functional. The optimized geometry was compared to crystallographic data in the Cambridge Crystallographic Data Center, such a comparison between the experimental and theoretical values helps to reduce the error in the optimized geometry. The optimized geometry was used for the calculations of harmonic vibrational frequencies at the B3LYP/6311G++(d,p) method, it also helps to ensure the systems to be local minimum number imaginary vibration frequencies. The thermochemical properties [10–12] like, hardness (η), softness (S), chemical potential (μ), electronegativity (χ) and electrophilicity index (ω), were calculated using Koopmans' theorem for closed-shell compounds. Electrostatic potential analysis has also been made to identify the mapping surface of drugs. Dipole moments, linear and non-linear optical (NLO) properties of AMB were also calculated at the same level of theory.

Further, molecular docking was also conducted to predict binding poses, bio affinity and virtual screening of the selected drugs into the 3D crystal structure of cyclooxygenase-2 (PDB ID: 1CX2) and cyclooxygenase-1 (PDB ID: 1EQG) using GLIDE Dock Program in Schrödinger Maestro software. The protein structure was refined using the protein preparation wizard, which employs under restrained minimization and heavy atoms were restrained by using OPLS 2003 force field. The ligands were subjected to ligand preparation using the ligand preparation wizard (Lig prep) of Schrödinger software in the Maestro interface (11.5). Grid center is defined for the active site and box sizes are set to 20 Å [2, 13, 14].

3. Results and discussion

3.1 Molecular geometry

The optimized structures of (a) fenoprofen, (b) ketoprofen, (c) flurbiprofen and (d) ibuprofen were calculated using B3LYP/6311G++(d,p) level of the theory and shown in **Figure 2**. The optimized geometries were compared to crystallographic data in the Cambridge Crystallographic Data Center to correlation coefficient factor. The calculated Pearson correlation coefficient for ketoprofen, fenoprofen and Ibuprofen has given in **Table 1**; the crystallographic data for flurbiprofen is not available yet. The Pearson correlation coefficient (PCC), or Pearson's r, the Pearson product-moment correlation coefficient (PPMCC), or the bivariate correlation is a statistic that measures linear correlation between two variables X and Y. Here, this method is used to find out the linear regression between the experimental and computationally calculated geometric parameters. Normally, it has a value between +1 and −1, where +1 indicates total positive linear correlation, 0 is no linear correlation, and −1 is total negative linear correlation [15].

3.2 Thermo-chemical properties

The thermo-chemical parameters, such as enthalpy (H), entropy (S), Gibb's free energy (G) were calculated to find which drug is more stable by comparing G and S

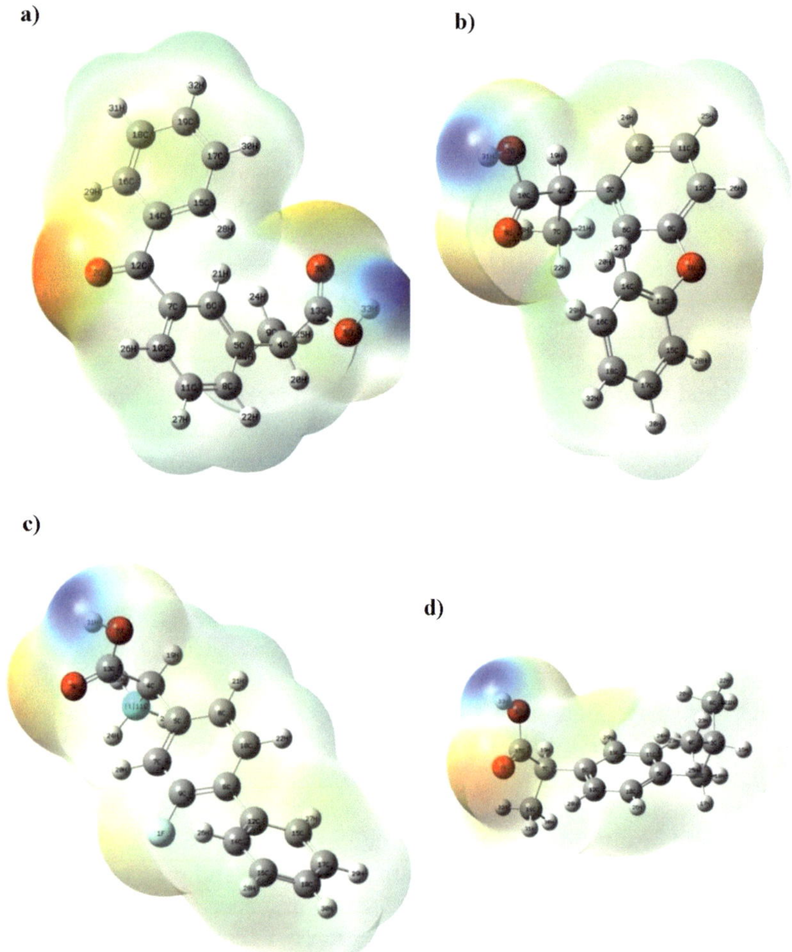

Figure 2.
The optimized molecular structures of the selected drugs, calculated at DFT/B3LYP/6311G++(d,p). (a) Ketoprofen. (b) Fenoprofen. (c) Flurbiprofen. (d) Ibuprofen.

Sample	Pearson correlation coefficient
Ketoprofen	0.951
Fenoprofen	0.976
Ibuprofen	0.899

Table 1.
The Pearson coefficient between the experimental and computationally calculated geometric parameters.

Sample	Energy $\times 10^5$ (kcal/mol)	Enthalpy $\times 10^5$ (kcal/mol)	Gibbs free energy $\times 10^5$ (kcal/mol)	Entropy (cal/mol)	Molecular mass (amu)
ketoprofen	−5.29	−5.29	−5.29	136.08	254.09
fenoprofen	−5.06	−5.06	−5.06	133.11	242.09
flurbiprofen	−5.21	−5.21	−5.21	129.43	244.08
ibuprofen	−4.12	−4.12	−4.12	128.56	206.13

Table 2.
Thermo-chemical properties of the selected drugs were calculated using the DFT/B3LYP/6311G++(d,p) level of theory.

SAMPLE	HOMO (eV)	LUMO (eV)	HOMO-1 (eV)	LUMO+1 (eV)	HOMO-2 (eV)	LUMO+2 (eV)	BAND GAP (eV)
Ketoprofen	−6.96	−2.10	−7.25	−1.02	−7.29	−0.88	4.86
Fenoprofen	−6.26	−0.91	−7.01	−0.65	−7.10	−0.48	5.36
Flurbiprofen	−6.54	−1.35	−7.15	−0.71	−7.21	−0.53	5.19
Ibuprofen	−6.68	−0.77	−7.05	−0.50	−7.92	−0.34	5.90

Table 3.
Frontier molecular orbital of the selected NSAIDs was calculated using the DFT/B3LYP/6311G++(d,p) level of theory.

values and the obtained values are given in **Table 2**. In general, more negative the value of G the drug is more stable and more active if the value of S is more positive. From the analysis, it is found that ketoprofen has more negative value for G (-5.29×10^5 kcal/mol) and flurbiprofen has more or less same G value (-5.21×10^5 kcal/mol). Hence, ketoprofen is a more stable and active drug compared to other selected drugs with enhanced entropy value of 136.08 cal/mol.

3.3 Frontier molecular orbital analysis

In computational chemistry, the frontier molecular orbitals play an important role in demonstrating active sites, kinetic stability and chemical reactivity of the molecule (**Table 3**). In the present work, frontier molecular orbital energies (E_{HOMO} and E_{LUMO}) of all selected drugs were calculated using DFT/B3LYP/6311G++(d,p) level of theory. The LUMO indicates the most likely site which would undergo a nucleophilic attack while the HOMO describes the most likely site for an electrophilic attack. The energy corresponding to HOMO represents the ionization potential of the molecule, while that of the LUMO represents the corresponding electron affinity. A high HOMO–LUMO energy gap indicates greater stability and low reactivity of the chemical system. On the basis of frontier molecular orbital analysis, ketoprofen is found to be more reactive with lower stability compared to other drug molecules in the family.

To understand the three-dimensional charge distributions over the drug molecules, to locate the most electronegative and electropositive site on their skeleton and to predict reactive sites for electrophilic and nucleophilic attack for the NSAIDs molecular electrostatic potential (MESP) mapping can sightsee.

The MESP map of ketoprofen shows that the negative potential sites are on electronegative atoms like oxygen atoms as well as the positive potential sites are around the hydrogen atoms. These sites give information about the region from where the compound can have noncovalent interactions (**Figure 3**).

3.4 Global descriptive parameters

In order to have a deep insight about the reactive nature of selected NSAIDs, the global descriptive parameters like hardness, softness, chemical potential, electronegativity, and electrophilicity index were calculated using Koopmans' theorem for closed-shell compounds, as follows [11, 16]:

$$\text{Ionization potential (IP)} \approx -E_{HOMO} \quad (1)$$

$$\text{Electron affinity (EA)} \approx -E_{LUMO} \quad (2)$$

where E_{HOMO} is the energy of HOMO and E_{LUMO} is the energy of LUMO.

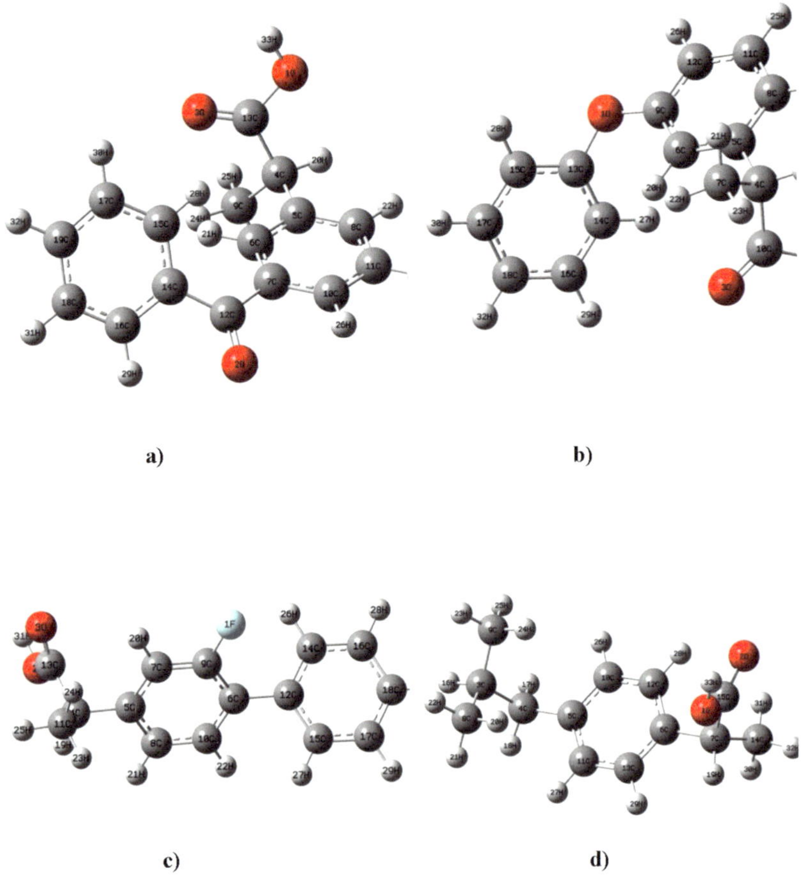

Figure 3.
The MESP structures of (a) ketoprofen, (b) fenoprofen, (c) flurbiprofen and (d) ibuprofen. The negative (red) regions of MESP were related to electrophilic reactivity and the positive (blue) regions to nucleophilic reactivity.

$$\text{Hardness } (\eta) \approx \frac{IP - EA}{2} \tag{3}$$

$$\text{Electronegativity } (\chi) \approx \frac{IP + EA}{2} \tag{4}$$

$$\text{Softness (S)} \approx \frac{1}{2\eta} \tag{5}$$

$$\text{Chemical potential } (\mu) \approx -\chi \tag{6}$$

$$\text{Electrophilicity index } (\omega) \approx \frac{\mu^2}{2\eta} \tag{7}$$

The calculated global descriptors of AMB are given in **Table 4.**

According to the maximum hardness principle (MHP), at constant external potential, the stability of a molecule increases with hardness, and with the increase in stability, the reactivity decreases. Softness is just the reciprocal of hardness, so higher the softness, lower is the stability, *i.e.*, higher is the reactivity. The hardness value of ketoprofen is 2.43; fenoprofen is 2.68, flurbiprofen is 2.60, and ibuprofen is 2.96. This study shows that ketoprofen has a lower hardness and a higher softness value which indicates that this drug is highly reactive compared to other drugs.

Sample	Ionization potential (Ip)	Electron affinity (Ea)	Hardness (η)	Electronegativity (χ)	Softness (S)	Chemical potential (μ)	Electrophilicity index (ω)
Ketoprofen	6.96	2.100	2.43	4.53	4.86	−4.53	4.86
Fenoprofen	6.26	0.90	2.68	3.58	5.36	−3.58	5.36
Flurbiprofen	6.54	1.35	2.60	3.95	5.19	−3.95	5.19
Ibuprofen	6.68	0.77	2.96	3.73	5.91	−3.73	5.91

Table 4.
Calculated global descriptors of the selected NSAIDs were calculated using the DFT/B3LYP/6311G++(d,p) level of theory.

Ionization energy is a fundamental descriptor of the chemical reactivity of atoms and molecules. High ionization energy indicates high stability and chemical inertness, and small ionization energy indicates high reactivity of the atoms and molecules. If the electronic chemical potential is greater, then the compound is less stable or more reactive. Electron affinity refers to the capability of a ligand to accept precisely one electron from a donor. The electrophilicity index is described as a structural depictor for the analysis of the chemical reactivity of molecules. It measures the tendency of the species to accept electrons. A good, more reactive, nucleophile has a lower value of ω, in the opposite, a good electrophile has a high value of ω. Hence comparing the local descriptors of all selected drug we can infer that ketoprofen is more reactive with the lowest electrophilicity index and highest softness index and smallest hardness value. At the same time, the negative chemical potential of ketoprofen determines the stability of the drugs.

3.5 Hyperpolarizabilities, polarisabilities and dipole moment

The computational approach can also be used to study the interaction of electromagnetic fields in various media to produce new fields that are altered in frequency, phase and amplitude or other propagation characteristics from the incident fields. The polarization P, induced in a medium by an external electric field F is given by

$$P = P_0 + \chi^{(1)} F + \chi^{(2)} F^2 + \chi^{(3)} F^3 + \ldots \tag{8}$$

where $\chi^{(n)}$ is the nth order susceptibility tensor of the bulk medium.

The dipole moment of a molecule interacting with an electric field can be written

$$\mu_i = \mu_i^0 + \alpha_{ij} F_j + (1/2)\beta_{ijk} F_j F_k + (1/6)\Upsilon_{ijkl} F_j F_k F_l + \tag{9}$$

where μ_i^0 is the permanent dipole moment and α_{ij}, β_{ijk}, Υ_{ijkl} is tensor elements of the linear polarizability and first and second hyperpolarizabilities respectively. This interaction may even lead to nonlinear optical effects (NLO). In this direction in order to study the NLO properties, the dipole moment, first static hyperpolarizability (βtot) and its related properties including α, β and Δα of all selected NSAIDs were calculated using DFT/B3LYP/6311G++(d,p) method based on the finite-field approach and are given in **Table 4**. The second-order term of the hyperpolarizability gives rise to sum and difference frequency mixing (including second harmonic generation) and optical rectification. The third-order term is responsible for the third-harmonic generation and two-photon resonances. The polarizability and hyperpolarizability of NLO can be written as tensors. While the linear polarizability tensor α as shown below which is a 3*3 matrix having nine components as shown below.

$$\alpha = \begin{bmatrix} \alpha_{xx} & \alpha_{xy} & \alpha_{xz} \\ \alpha_{yx} & \alpha_{yy} & \alpha_{yz} \\ \alpha_{zx} & \alpha_{zy} & \alpha_{zx} \end{bmatrix} \tag{10}$$

$$\alpha_{total} = \frac{(\alpha_{xx} + \alpha_{yy} + \alpha_{zz})}{3} \tag{11}$$

For the first hyperpolarizability, the quantity of interest β is a 3*3*3 matrix has β_{xzz}, β_{xxx}, β_{xyy}, β_{yyy}, β_{yzz}, β_{yxx}, β_{zzz}, β_{zxx} β_{zyy}, β_{xyz}, respectively, from which the x, y and z components of β are calculated as [12, 17–27]:

$$\beta_{total} = \sqrt{\left(\beta_{xxx} + \beta_{xyy} + \beta_{xzz}\right)^2 + \left(\beta_{yyy} + \beta_{yzz} + \beta_{yxx}\right)^2 + \left(\beta_{zzz} + \beta_{zxx} + \beta_{zyy}\right)^2} \quad (12)$$

The highest value of dipole moment is observed for ketoprofen, which is equal to 3.2078 Debye. The calculated average polarizability and first-order hyperpolarizability of the drug molecules are given in **Table 5**. All the samples exhibit better values compared to one of the prototypical molecules, Urea (μ and β of urea is 4.56 D and 4.8×10^{-36} esu respectively) [25]. Though all the molecules are NLO active molecule, ketoprofen has the highest among others with hyperpolarizability value of 9.2128×10^{-31} esu.

3.6 Molecular docking

Molecular docking is one of the most frequently used methods in structure-based drug design due to its ability to predict the binding conformation of small molecules to the appropriate target binding site.

3.6.1 Molecular docking studies with COX-2

1CX2 is an enzyme involved in arachidonic acid metabolism, where arachidonic acid is the precursor that is metabolized by various enzymes, especially cyclooxygenase-1 and -2 to a wide range of biologically and clinically important eicosanoids and metabolites of these eicosanoids. Two important pathways for arachidonic acid metabolism are the cyclooxygenase (COX) and 5-lipoxygenase (5-LO) pathways. The COX pathway forms intermediate compounds called cyclo-endoperoxides (PGG2 and PGH2). Enzymes, many of which are tissue specific, then convert the cyclo-endoperoxides into the final biologically active prostanoid. 1CX2 is a tetramer having four identical subunits and its chains C and D contain the inbuilt ligand SC-558 and the structure of the COX-2 enzyme is depicted in **Figure 4**. The amino acid residues in the active site of 1CX2 are TYR355, GLY354, SER353, LEU352, VAL349, TYR348, MET522, VAL523, GLY526, ALA527, SER530, LEU531, LEU359, ARG120, VAL116, HIS90, ARG513, ALA516, ILE517, PHE518, TRP387, TYR385, LEU384 and PHE381.

Figure 5(b) demonstrates the 3-dimensional protein-ligand interaction SC-558 into the active site of 1CX2. The co-crystallized ligand SC-558 is found to be buried deep into the binding pocket of 1CX2. SC-558 interacts with the active site's amino acids of the protein by H-bonding with active site amino acids ARG513, SER353 and LEU352, which is well evidently observed from 2-D interaction picture as shown in **Figure 5(c)**. SC-558 interacts with COX-2 with a binding energy of −11.987 kcal/mol.

After analyzing the protein-ligand interaction of 1CX2 with its co-crystallized ligand SC558, all the selected ligands ketoprofen, fenoprofen, flurbiprofen and ibuprofen were docked to the active site of 1CX2 and it is found that all the selected

Samples	Dipole moment (μ) Debye	Polarizability (esu)	Hyperpolarizability (esu)
Ketoprofen	3.20	-1.59×10^{-23}	9.21×10^{-31}
Ibuprofen	1.70	-1.35×10^{-23}	4.55×10^{-31}
Fenoprofen	1.16	-1.50×10^{-23}	5.17×10^{-31}
Flurbiprofen	2.68	-1.50×10^{-23}	3.95×10^{-31}

Table 5.
The calculated dipole moment μ (Debye), the polarizability β_{tot} and first hyperpolarizability β_{tot} of all selected NSAIDs.

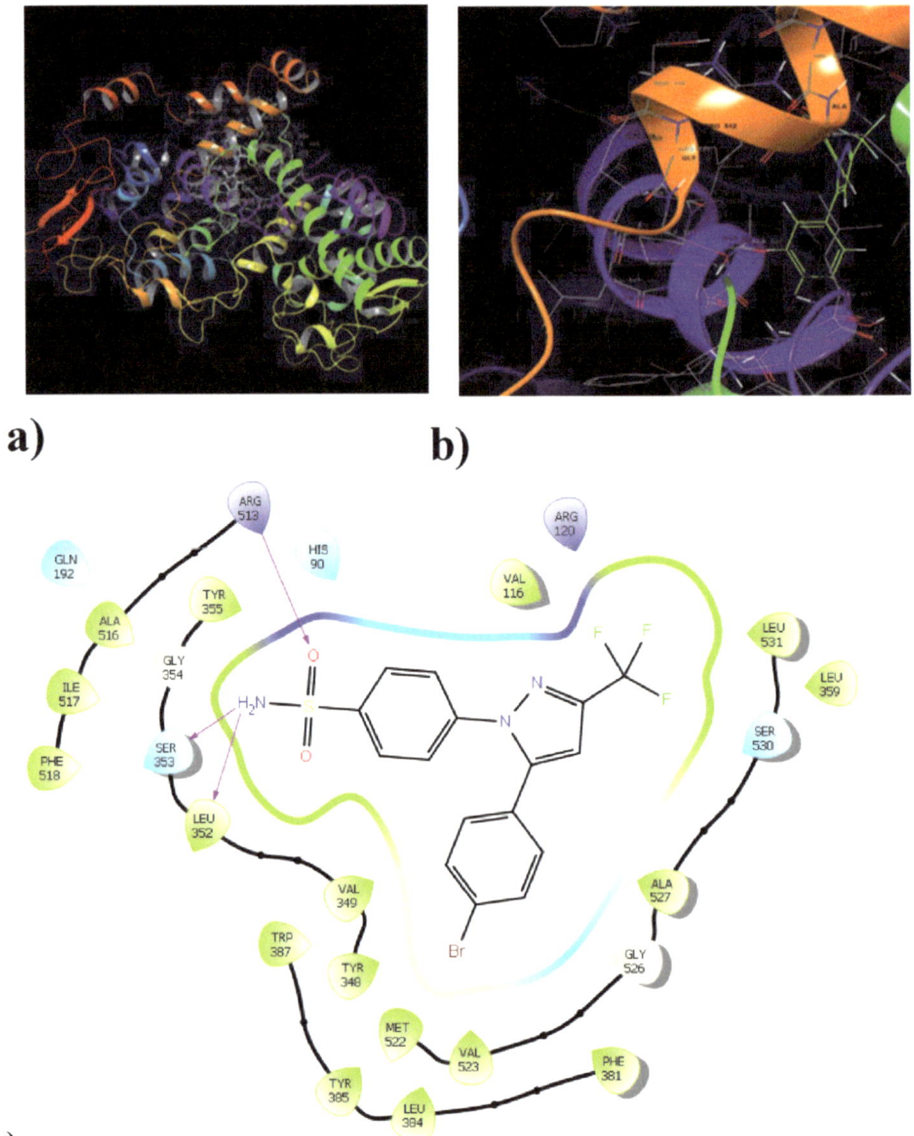

Figure 4.
(a) Structure of the COX-2 enzyme having inbuilt ligand SC-558. (b) Protein-ligand interaction of SC-558 with 1CX2. (c) 2-D protein-ligand interaction of SC-558 with 1CX2.

NSAIDs docked well deep into the binding pocket of 1CX2 with good gliding score, given in **Table 5**. **Figure 5** shows the 3-D protein-ligand interaction of all selected NSAIDs into the active site of COX-2.

Ketoprofen and fenoprofen were docked deeply into the active site region making interactions with the residues ARG120, TYR355, TYR385 and TRP387 while, flurbiprofen docked deeply into the active site region making interactions with the residues ARG120 and TYR355 and ibuprofen with residue TYR355 only (**Figure 6**).

3.6.2 Molecular docking studies with COX-1

COX-1 is an enzyme that acts on arachidonic acid and produces housekeeping prostaglandins. It is a dimer having two identical structural units in which Chain B

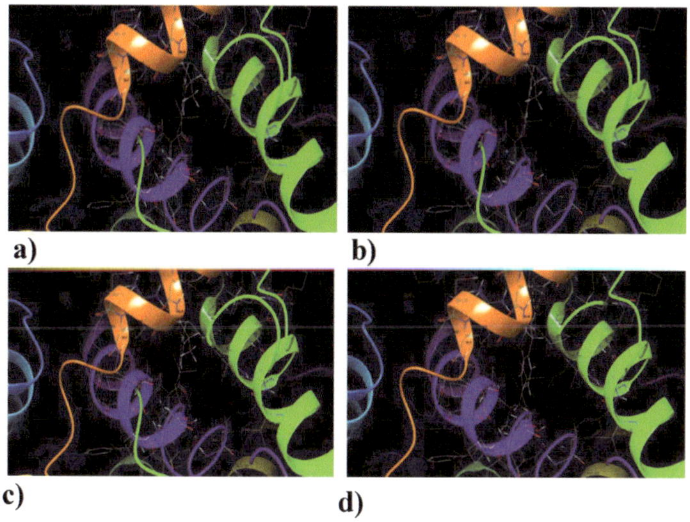

Figure 5.
3-D protein-ligand interactions of the ligands (a) ketoprofen, (b) fenoprofen, (c) flurbiprofen and (d) ibuprofen into the active site of COX-2.

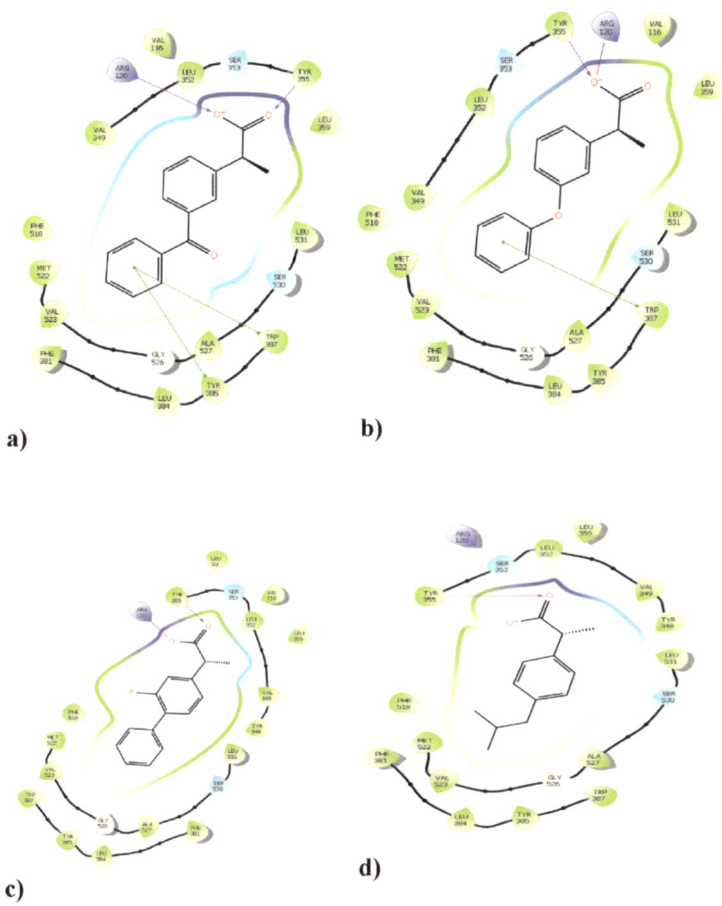

Figure 6.
2-D protein-ligand interactions of the ligands: (a) ketoprofen, (b) fenoprofen, (c) flurbiprofen and (d) ibuprofen into the active site of COX-2.

and some nonstandard residues were deleted after the preprocessing and the structure of the protein are shown in **Figure 7**. The amino acid residues in the active site of 1EQG are TYR355, SER353, LEU352, VAL349, MET522, ILE523, ALA527, SER530, LEU531, LEU359, ARG120, VAL116, PHE518, GLY526, MET522, PHE518, TRP387, TYR385, LEU384 and PHE381.

All the selected ligands ketoprofen, fenoprofen, flurbiprofen, and ibuprofen were docked to the active site of COX-1 and it is found that all the selected NSAIDs docked well deep into the binding pocket of COX-1 with good gliding score, given in **Table 6** and binding energy were tabulated in **Table 7**. **Figure 8** shows the 3-D protein-ligand interaction of all selected NSAIDs into the active site of COX-1.

Ketoprofen and fenoprofen were docked deeply into the active site region making interactions with the residues TYR385, TRP387, ARG120 and TYR355 by forming two hydrogen bonds with ARG120 and TYR355, two pi-pi stacking

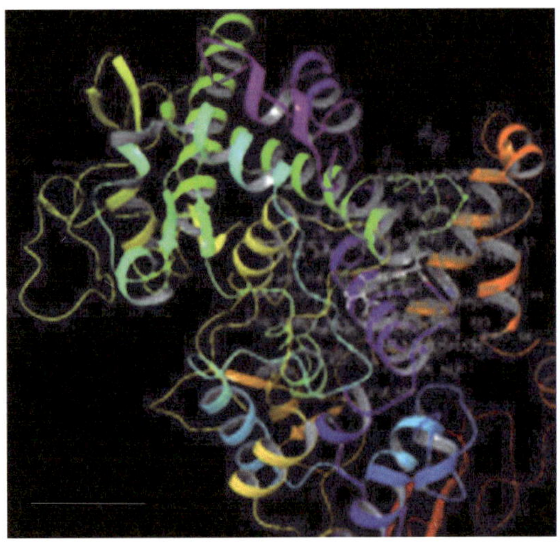

Figure 7.
Structure of COX-1.

Sample	Gliding score with COX-2	Gliding score with COX-1
Ketoprofen	−9.279	−11.242
Fenoprofen	−11.37	−10.862
Flurbiprofen	−9.377	−11.602
Ibuprofen	−7.468	−10.075

Table 6.
The gliding score of all samples with COX-1 and COX-2 enzymes.

Sample	Binding energy with COX-2 kcal/mol	Binding energy with COX-1 kcal/mol
Ketoprofen	−9.280	−11.242
Fenoprofen	−8.694	−10.863
Flurbiprofen	−9.377	−11.603
Ibuprofen	−10.133	−10.075

Table 7.
The binding energy of all samples with COX-1 and COX-2 enzymes.

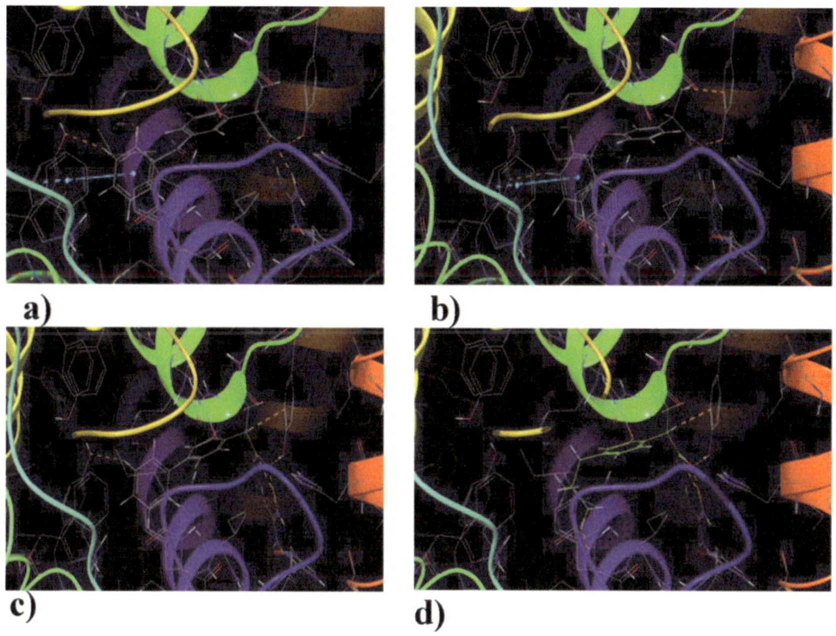

Figure 8.
3-D protein-ligand interactions of the ligands (a) ketoprofen, (b) fenoprofen, (c) flurbiprofen, (d) ibuprofen into the active site of COX-1.

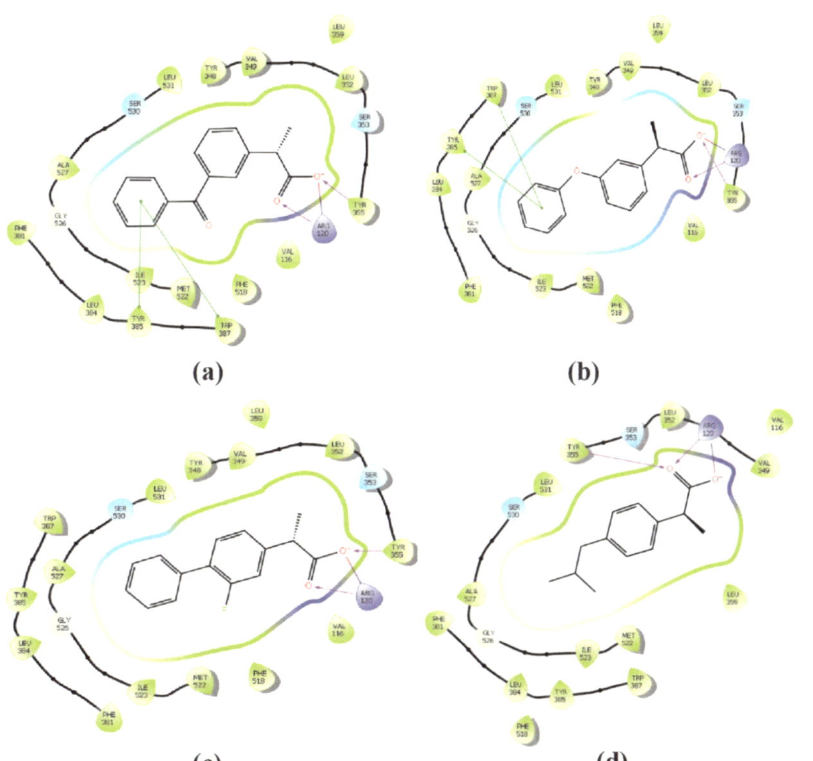

Figure 9.
2-D protein-ligand interactions of the ligands (a) ketoprofen, (b) fenoprofen, (c) flurbiprofen and (d) ibuprofen into the active site of COX-1.

interaction between phenyl ring of the ligand with amino acids TYR385 and TRP387 and one salt bridge with active site amino acids. Though the interactions are the same for ketoprofen and fenoprofen the binding energy is different, ketoprofen has a binding energy of −11.242 kcal/mol while that of fenoprofen is −10.863 kcal/mol. At the same time, flurbiprofen and ibuprofen docked deeply into the active site region making interactions with the residues ARG120 and TYR355 and ibuprofen with residue TYR355 only (**Figure 9**).

4. Conclusion

The complete Quantum computational investigations were done using DFT theoretical calculation at the DFT/B3LYP/6311G++(d,p) method that has been performed for all selected propionic acid derivatives, NSAIDs. Almost all bond lengths and angles all the profen drugs agree very well with the X-ray crystal structures in Cambridge Crystallographic Data Center suggesting that all the molecules are well described with DFT/B3LYP/6311G++(d,p) level of theory. The electrophilic and nucleophilic sites were traced out from the isosurfaces of molecular electrostatic potential. The detailed confab on the calculated global descriptors revealed that ketoprofen is more reactive than other propionic derivatives and has the ability to donate electrons easily. Though the hyperpolarizability values reveal that the all selected organic molecule has better NLO activity compared to urea, ketoprofen shows better activity than others. Further, the molecular docking studies of these compounds demonstrates a good selectivity profile with both COX enzymes with good gliding scores and confirmed ketoprofen is a strong anti-inflammatory agent compared to others.

Acknowledgements

Authors thankfully acknowledge the Central Sophisticated Instrument Facility (CSIF) of University of Calicut for providing Schrodinger Maestro software and hardware support. KPSH further acknowledges UGC-MANF for fellowship with sanction number MANF2017-18-KER-78598.

Conflict of interest

The authors declare no conflict of interest.

Author details

Safna Hussan Kodakkat Parambil[1], Hisana Asharaf Thozhuvana Parambil[2], Shahina Parammal Hamza[3], Anjali Thirumangalath Parameswaran[3], Mohamed Shahin Thayyil[1] and Muraleedharan Karuvanthodi[4*]

1 Department of Physics, University of Calicut, Malappuram, India

2 Department of Chemistry, St. Mary's College Sulthan Bathery, Kerala, India

3 Department of Chemistry, Little Flower College, Guruvayur, Kerala, India

4 Department of Chemistry, University of Calicut, India

*Address all correspondence to: kmuralika@gmail.com

References

[1] S. Asirvatham, B. V. Dhokchawle, S.J. Tauro, Quantitative structure activity relationships studies of non-steroidal anti-inflammatory drugs: A review, Arab. J. Chem. 12 (2016) 3948–3962. doi:10.1016/j.arabjc.2016.03.002.

[2] S. Aktar, M.F. Khan, M.M. Rahman, M.A. Rashid, Computational study of geometry, polarizability, hyperpolarizability and molecular docking studies of naproxen, Dhaka Univ. J. Pharm. Sci. 15 (2016) 37–45. doi: 10.3329/dujps.v15i1.29191.

[3] N. Okulik, A.H. Jubert, Theoretical study on the structure and reactive sites of three non-steroidal anti-inflammatory drugs: Ibuprofen, Naproxen and Tolmetin acids, J. Mol. Struct. THEOCHEM. 769 (2006) 135–141. doi:10.1016/j.theochem.2005.10.061.

[4] P. Swiatek, K. Gebczak, T. Gebarowski, R. Urniaz, Biological evaluation and molecular docking studies of dimethylpyridine derivatives, Molecules. 24 (2019) 1–10. doi:10.3390/molecules24061093.

[5] A. Khalid, SaimaKalsoom, NaveedaRiaz, Design and Molecular Docking of Antioxidant Lead Compound and its Analogues Acting as Human Tyrosine Kinase Inhibitors, J. Pharm. Biol. Sci. 5 (2013) 75–80.

[6] Y.S. Mary, Y.S. Mary, K.S. Resmi, R. Thomas, DFT and molecular docking investigations of oxicam derivatives, Heliyon. 5 (2019) e02175. doi:10.1016/j.heliyon.2019.e02175

[7] E.E. Bolton, Y. Wang, P.A. Thiessen, S.H. Bryant, Integrated Platform of Small Molecules and Biological Activities, in: R.A.W. and D.C.S.B.T.-A. R. in C. Chemistry (Ed.), Elsevier, 2008: pp. 217–241. doi:10.1016/S1574-1400(08)00012-1.

[8] N.M. O'Boyle, M. Banck, C.A. James, C. Morley, T. Vandermeersch, G.R. Hutchison, Open Babel: An open chemical toolbox. J. Chem. Informatics. 3 (2011) 33

[9] M.J. Frisch, G.W. Trucks, H.B. Schlegel, G.E. Scuseria, M.A. Robb, C. J. R., G. Scalmani, V. Barone, B. Mennucci, G.A. Petersson, H. Nakatsuji, M. Caricato, X. Li, H.P. Hratchian, A.F. Izmaylov, J. Bloino, G. Zheng, J.L. Sonnenberg, M. Hada, M. Ehara, K. Toyota, R. Fukuda, J. Hasegawa, M. Ishida, T. Nakajima, Y. Honda, O. Kitao, H. Nakai, E. T. Vreven, J.A. Montgomery, Jr., J.E. Peralta, F. Ogliaro, M. Bearpark, J.J. Heyd, A. Brothers, K. N. Kudin, V.N. Staroverov, R. Kobayashi, J. Normand, K. Raghavachari, M.K. Rendell, J.C. Burant, S.S. Iyengar, J. Tomasi, M. Cossi, N. Rega, J.M. Millam, R.E.S. J.E. Knox, J.B. Cross, V. Bakken, C. Adamo, J. Jaramillo, R. Gomperts, K.M. O. Yazyev, A.J. Austin, R. Cammi, C. Pomelli, J. Ochterski, R.L. Martin, O. V. G. Zakrzewski, G.A. Voth, P. Salvador, J.J. Dannenberg, S. Dapprich, A.D. Daniels, D.F. Farkas, J.B. Foresman, J. V. Ortiz, J. Cioslowski, J., GAUSSIAN 09 (Revision A.2) Gaussian, Inc., Wallingford, CT, D. J. Gaussian 09,. B.01; (1998).

[10] V.K. Rajan, K. Muraleedharan, A computational investigation on the structure, global parameters and antioxidant capacity of a polyphenol, Gallic acid, Food Chem. 220 (2017) 93–99. doi:10.1016/j.foodchem.2016.09.178.

[11] K.P. Safna Hussan, M.S. Thayyil, V. K. Rajan, K. Muraleedharan, Experimental and density functional theory studies on benzalkonium ibuprofenate, a double active pharmaceutical ingredient, Comput. Biol. Chem. 72 (2018) 113–121. doi: 10.1016/j.compbiolchem.2017.12.004.

[12] K. Safna Hussan, M.S. Thayyil, V.K. Rajan, K. Muraleedharan, DFT studies on global parameters, antioxidant mechanism and molecular docking of amlodipine besylate, Comput. Biol. Chem. 80 (2019) 46–53. doi:10.1016/j.compbiolchem.2019.03.006.

[13] M. Szermerski, J. Melesina, K. Wichapong, M. Löppenberg, J. Jose, W. Sippl, R. Holl, Synthesis, biological evaluation and molecular docking studies of benzyloxyacetohydroxamic acids as LpxC inhibitors, Bioorg. Med. Chem. 22 (2014) 1016–1028. doi:10.1016/j.bmc.2013.12.057.

[14] T. Schulz-Gasch, M. Stahl, Binding site characteristics in structure-based virtual screening: evaluation of current docking tools, J. Mol. Model. 9 (2003) 47–57. doi:10.1007/s00894-002-0112-y.

[15] Pearson correlation coefficient, (n. d.). https://en.wikipedia.org/wiki/Pearson_correlation_coefficient.

[16] K.G. Sangeetha, K.K. Aravindakshan, K.P. Safna Hussan, Insight into the theoretical and experimental studies of 1-phenyl-3-methyl-4-benzoyl-5-pyrazolone N (4)-methyl- N (4)-phenylthiosemicarbazone - A potential NLO material, J. Mol. Struct. 1150 (2017) 135–145. doi:10.1016/j.molstruc.2017.08.078.

[17] A. Kumar, V. Deval, P. Tandon, A. Gupta, E.D. Deepak, Molecular and Biomolecular Spectroscopy Experimental and theoretical (FT-IR, FT-Raman, UV – vis, NMR) spectroscopic analysis and first order hyperpolarizability studies phenyl] -1-(4-nitrophenyl) prop-2-en-1-one using, Spectrochim. Acta Part A Mol. Biomol. Spectrosc. 130 (2014) 41–53. doi:10.1016/j.saa.2014.03.072.

[18] C. Tabti, N. Benhalima, Molecular Structure, Vibrational Assignments and Non-Linear Optical Properties of by DFT and ab Initio HF Calculations, Adv. Mater. Phys. Chem. 5 (2015) 221–228.

[19] G. Ramachandran, S. Muthu, J. Uma Maheswari, Density functional theory and Ab initio studies of vibrational spectroscopic (FT-IR, FT-Raman and UV) first order hyperpolarizabilities, NBO, HOMO-LUMO and TD-DFT analysis of the 1,2-Dihydropyrazolo (4,3-E) Pyrimidin-4-one, Solid State Sci. 16 (2013) 45–52. doi:10.1016/j.solidstatesciences.2012.11.005.

[20] E. Temel, C. Alaşalvar, H. Gökçe, A. Güder, Ç. Albayrak, Y.B. Alpaslan, G. Alpaslan, N. Dilek, DFT calculations, spectroscopy and antioxidant activity studies on (E)-2-nitro-4-[(phenylimino)methyl]phenol, Spectrochim. Acta - Part A Mol. Biomol. Spectrosc. 136 (2015) 534–546. doi:10.1016/j.saa.2014.09.067.

[21] T. Rajamani, S. Muthu, Electronic absorption, vibrational spectra, non-linear optical properties, NBO analysis and thermodynamic properties of 9-[(2-hydroxyethoxy) methyl] guanine molecule by density functional method, Solid State Sci. 16 (2013) 90–101. doi:10.1016/j.solidstatesciences.2012.10.023.

[22] K.P. Safna Hussan, T. Mohamed Shahin, S.K. Deshpande, Studies of Ionogel Structure and its Electronic and Optical Characterization by ONIOM and other Hybrid Computational Approaches, Mater. Today Proc. 5 (2018) 16272–16279. doi:10.1016/j.matpr.2018.05.119.

[23] E.B. Sas, E. Kose, M. Kurt, M. Karabacak, FT-IR, FT-Raman, NMR and UV-Vis spectra and DFT calculations of 5-bromo-2-ethoxyphenylboronic acid (monomer and dimer structures), Spectrochim. Acta - Part A Mol. Biomol. Spectrosc. 137 (2015) 1315–1333. doi:10.1016/j.saa.2014.08.049.

[24] A. Kumar, V. Deval, P. Tandon, A. Gupta, E.D. D'Silva, Experimental and

theoretical (FT-IR, FT-Raman, UV-vis, NMR) spectroscopic analysis and first order hyperpolarizability studies of non-linear optical material: (2E)-3-[4-(methylsulfanyl) phenyl]-1-(4-nitrophenyl) prop-2-en-1-one using density functional th, Spectrochim. Acta - Part A Mol. Biomol. Spectrosc. 130 (2014) 41–53. doi:10.1016/j.saa.2014.03.072.

[25] S. Muthu, E. Elamurugu Porchelvi, FTiR, FT-RAMAN, NMR, spectra, normal co-ordinate analysis, NBO, NLO and DFT calculation of N, N-diethyl-4-methylpiperazine-1-carboxamide molecule, Spectrochim. Acta - Part A Mol. Biomol. Spectrosc. 115 (2013) 275–286. doi:10.1016/j.saa.2013.06.011.

[26] K.P.S. Hussan, M.S. Thayyil, Detailed Quantum mechanical Investigation on Global descriptors, NBO , NLO, radical scavenging activities and Molecular Docking of a dihydropyridine calcium channel blocker – amlodipine besylate, (n.d.) 311.

[27] S. Xavier, S. Periandy, K. Carthigayan, S. Sebastian, Molecular docking, TG/DTA, molecular structure, harmonic vibrational frequencies, natural bond orbital and TD-DFT analysis of diphenyl carbonate by DFT approach, J. Mol. Struct. 1125 (2016) 204–216. doi:10.1016/j.molstruc.2016.06.071.

Section 6

Non-equilibrium Equation and Density Functional

Chapter 6

Non-equilibrium Equation of State in the Approximation of the Local Density Functional and Its Application to the Emission of High-Energy Particles in Collisions of Heavy Ions

A.T. D'yachenko and I.A. Mitropolsky

Abstract

The non-equilibrium equation of state is found in the approximation of the functional on the local density, and its application to the description of the emission of protons and pions in heavy ion collisions is considered. The non-equilibrium equation of state is studied in the context of the hydrodynamic approach. The compression stage, the expansion stage, and the freeze-out stage of the hot spot formed during the collisions of heavy ions are considered. The energy spectra of protons and subthreshold pions produced in collisions of heavy ions are calculated with inclusion of the nuclear viscosity effects and compared with experimental data for various combinations of colliding nuclei with energies of several tens of MeV per nucleon.

Keywords: local density functional, hydrodynamics, non-equilibrium equation of state, heavy ions, hot spot, nuclear viscosity, protons, subthreshold pions

1. Introduction

The main object of studying heavy ion collisions is to study the equation of state (EOS) of nuclear matter. Along with molecular dynamics and the Vlasov dynamic equation, nuclear hydrodynamics is an effective method for describing the interaction of heavy ions with medium and intermediate energies (see, e.g., [1]). Typically, the equilibrium EOS is used [1]; it involves the local thermodynamic equilibrium in the system. A hybrid model was proposed for use at high energies in [2, 3]. It includes a fast non-equilibrium stage and the subsequent description of the dynamics of a nucleus-nucleus collision based on equilibrium relativistic hydrodynamics of an ideal fluid. We showed in our works [4–11] that local thermodynamic equilibrium is not immediately established in the process of collisions of heavy ions, since the non-equilibrium component of the distribution function, which leads to the formation of a collisionless shock wave, is important at the compression stage.

 IntechOpen

The kinetic equation for finding the distribution function of nucleons is used in this paper. It is solved in conjunction with the equations of hydrodynamics, which are essentially local conservation laws of mass, momentum, and energy. As a result, the non-equilibrium equation of state is found in the approximation of the functional on the local density. Since the emitted secondary particles (nucleons, fragments, and pions) contain the basic information about the EOS, it is necessary to know the differential cross sections for the emission of these particles. The energy spectra of protons and subthreshold pions with allowance for nuclear viscosity are analyzed in this paper as a follow-up to our works [11–13] devoted to the energy spectra of protons and fragments in which viscosity was neglected.

By subthreshold production, we mean the generation of π mesons with energies lower than the threshold for the production of pions E_{NN} in free nucleon-nucleon collisions. The absolute thresholds for pion production are $E_{NN} = 2m_\pi + \frac{m_\pi^2}{2m} \approx 290$ MeV in nucleon-nucleon collisions, $E_{NA} \approx m_\pi \approx 140$ MeV in nucleon-nucleus collisions, and $E_{BA} = \frac{m_\pi^2 + 2(A+B)m_\pi m}{2ABm} \approx 20$ MeV in nucleus-nucleus collisions at $A = B = 12$, where m_π is the pion mass and m is the nucleon mass. This expression for the absolute threshold energy is obtained from a comparison of the relativistic invariants $J = E^2 - P^2$ before and after the collision, neglecting the binding energy of pion (E is the total energy; P is the total momentum).

The pion production threshold during the collision of heavy ions decreases owing to collective effects and the internal motion of nucleons. These effects are naturally taken into account using the hydrodynamic approach, which explicitly includes the many-particle nature of colliding heavy ions. In the case of low energies, the hydrodynamics should be modified to take into account the non-equilibrium EOS, which describes the transition from the initial non-equilibrium state to the state of local thermodynamic equilibrium.

Such an approach to describing the temporal evolution of the resulting hot spot includes a compression stage and an expansion stage taking into account the nuclear viscosity that we found. The calculated energy spectra of protons and pions produced in nuclear collisions (both identical and different in mass) at an energy of 92 MeV per nucleon in the case of protons and 94 MeV per nucleon in the case of subthreshold pions are in agreement with the available experimental data [1, 14], respectively.

2. Non-equilibrium equation of state in a local density approximation

If the energies of colliding heavy ions are less than 300 MeV per nucleon (pion production threshold in free nucleon-nucleon collisions), we use the kinetic equation to find the nucleon distribution function $f\left(\vec{r}, \vec{p}, t\right)$ ($\vec{r}\,(x_1, x_2, x_3)$ is the spatial coordinate; $\vec{p}\,(p_1, p_2, p_3)$ is the momentum; t is the time) [11, 12]:

$$\frac{\partial f}{\partial t} + \frac{p_i}{m}\frac{\partial f}{\partial x_i} - \frac{\partial W}{\partial x_i}\frac{\partial f}{\partial p_i} = \frac{f_0 - f}{\tau}, \tag{1}$$

where $f_0\left(\vec{r}, \vec{p}, t\right)$ is a local equilibrium distribution function; τ is the relaxation time; $W(\rho)$ ($W(\rho) = \alpha\rho + \beta\rho^\gamma$) is a one-particle Skyrme-type self-consistent potential depending on the density ρ, where three parameters $\alpha(\alpha < 0)$, $\beta(\beta > 0)$, and $\gamma(\gamma > 1)$ are determined by setting the equilibrium density $\rho_0 = 0.145$ fm^{-3}, binding energy $E_b = -16$ MeV, and compression modulus $K = 210$ MeV; and m is the nucleon mass.

Equation (1) with allowance for the hydrodynamic equations obtained from (1) by taking the corresponding moments with a weight of 1, $\vec{p}$, and $\frac{p^2}{2m}$ [11, 12] describes the dynamics of nuclear collisions and forms the basis of our approach. The solution of Eq. (1) can be simplified if we work out the distribution function $f\left(\vec{r},\vec{p},t\right)$ determining EOS in the form

$$f\left(\vec{r},\vec{p},t\right)=f_1 q+f_0(1-q), \tag{2}$$

where the distribution function $f_1\left(\vec{r},\vec{p},t\right)$ is defined in momentum space as an axially symmetric Fermi ellipsoid, which is a convenient form for describing excitations in the Fermi liquid theory and is assumed to be blurred along the axis p_1 with the temperature T_1 and frozen in the transverse directions p_2 and p_3. The function $f_0\left(\vec{r},\vec{p},t\right)$ is represented in the momentum space by the equilibrium Fermi sphere blurred with temperature T; q is a relaxation factor ($q = \exp\left(-\int\limits_{t_0}^{t} dt/\tau\right)$), where $t_0\left(\vec{r},t\right)$ is the start time of the relaxation process in the system; τ is the relaxation time, which can be specified as in [15]. However, we define τ more traditionally as $\tau = \lambda/\upsilon_T$, where λ is the mean free path of nucleons at a given nucleon density, which is assumed to be equal to the mean distance between nucleons, and υ_T is the mean speed of the thermal Fermi motion of nucleons. This expression for τ in the energy range under consideration is close in magnitude to the value proposed in [15], but it turns out to depend on temperature and compression ratio and seems to us more realistic. All calculations are carried out precisely for such τ. The equation for finding the relaxation factor $q\left(\vec{r},t\right)$ is obtained by taking the moment for the kinetic equation with a weight of $p_1^2-\left(p_2^2+p_3^2\right)/2$ that determines the degree of anisotropy of the distribution function $f\left(\vec{r},\vec{p},t\right)$ in momentum space.

So, the initial moments $\int \frac{d^3\vec{p}}{(2\pi\hbar)^3}$ with weights 1, $\vec{p}$, $\frac{p^2}{2m}$, $p_1^2-\left(p_2^2+p_3^2\right)/2$ from kinetic equation (1) provide the corresponding hydrodynamic equations [11, 12] for finding nucleon density $\rho\left(\vec{r},t\right)$, velocity field $\vec{\upsilon}\left(\vec{r},t\right)$, internal energy density $e = \varepsilon + I + e_{int}$, and pressure tensor $P_{ij} = P_{(kin)ij} + P_{int}\delta_{ij}$ (the repeated indices imply the summation, δ_{ij} is the Kronecker symbol). The terms of interaction for energy density e_{int} and pressure P_{int} are, respectively,

$$e_{int} = \int\limits_0^{\rho} W(\rho)d\rho, P_{int} = \rho^2\frac{d(e_{int}/\rho)}{d\rho}.$$

The kinetic terms are.
$\varepsilon = \frac{3}{10}\frac{\hbar^2}{m}\left(\frac{3}{2}\pi^2\rho\right)^{2/3}\rho, \varepsilon_1 = \frac{\hbar^2}{10m}\left(\frac{3}{2}\pi^2\rho_0\right)^{2/3}\frac{\rho^3}{\rho_0^2}, \varepsilon_2 = \frac{\hbar^2}{10m}\left(\frac{3}{2}\pi^2\rho_0\right)^{2/3}\rho,$
$P_{kin}^{||} = P_{(kin)11} = 2(\varepsilon_1 + I_1)q + \frac{2}{3}(\varepsilon + I)(1-q),$
$P_{kin}^{\perp} = P_{(kin)22} = P_{(kin)33} = 2\varepsilon_2 q + \frac{2}{3}(\varepsilon + I)(1-q),$
which corresponds to diagonal tensor of pressure $P_{(kin)ij} = 0 (i \neq j)$, and heat terms I and I_1 are associated with temperatures T and T_1, respectively. Since we assume that the integrals of motion (density ρ, momentum density $m\rho\vec{\upsilon}$, and

energy density ($\varepsilon + I = \varepsilon_1 + 2\varepsilon_2 + I_1$)) are conserved during relaxation, hydrodynamic equations have no right-hand sides.

To find density ρ, velocity field $\vec{\upsilon}$ relaxation factor q, and temperatures T and T_1, we thus have the closed system of equations that considers expressions for terms of interaction and kinetic terms. These equations allow us to find distribution function $f\left(\vec{r}, \vec{p}, t\right)$ in form (2). Relaxation factor $q(0 \leq q \leq 1)$ allows us to describe the dynamics of the Fermi surface variation from the state with $q = 1$, where function $f\left(\vec{r}, \vec{p}, t\right)$ in the momentum space is maximally anisotropic, to the state with $q = 0$, where it is completely isotropic.

3. Hydrodynamic stage

We simplify the description of the time evolution of colliding nuclei distinguishing the compression stage, the expansion stage, and the freeze-out stage of the resulting hot spot. We reduce the interaction between two nuclei to the interaction between their overlapping regions. This can be interpreted as a hot spot formation process. In this case, we take into account the conservation laws. Shock waves with changing front diverging in opposite directions are formed at the stage of compression during the interaction between overlapping regions of colliding nuclei [5–9].

In the process of compression, when the shock wave reaches the boundaries of the hot spot, the density reaches its maximum value. The dependence of the maximum compression ratio ρ/ρ_0 at the shockwave front (solid line) on the collision energy of nuclei E_0 is shown in **Figure 1**. It hardly depends on the composition of colliding nuclei, since we consider the interaction of the same overlapping regions in the system of equal speeds of the colliding nuclei. The dependence of ρ/ρ_0 on the energy E_0 for the distribution function corresponding to the equilibrium EOS with $q = 0$ is shown by a dashed line, and such a dependence for a completely non-equilibrium EOS with $q = 1$ is shown by a dash-dotted line.

The relaxation factor at the energy region of $E_0 < 100$ MeV per nucleon is maximal ($q = 1$), and it decreases with increasing energy, leading to a greater isotropy of the distribution function. We calculated the dependence of the maximum compression ratio on energy for $E_0 > 100$ MeV per nucleon. It is found in between the extreme cases with $q = 0$ and $q = 1$. At $E_0 < 100$ MeV per nucleon, the dependence ρ/ρ_0 on energy coincides with the dash-dotted curve corresponding to the case with $q = 1$ (i.e., completely non-equilibrium EOS) and is located above the dashed curve corresponding to the case with $q = 0$ belonging to traditional hydrodynamics and the onset of local thermodynamic equilibrium.

A compressed and heated hot spot (a ball with radius R) expands when the shock wave reaches the boundaries of the system. The hot spot expands in accordance with the equations of hydrodynamics for radial motion of nucleon density $\rho(r,t)$, velocity $\upsilon(r,t)$, energy density $e(r,t)$, and pressure $P(r,t)$, following from (1) [11, 12]:

$$\frac{\partial \rho}{\partial t} + \frac{\partial (r^2 \rho \upsilon)}{r^2 \partial r} = 0, \tag{3}$$

$$\frac{\partial (m \rho \upsilon)}{\partial t} + \frac{\partial (r^2 m \rho \upsilon^2)}{r^2 \partial r} + \frac{\partial P}{\partial r} = 0, \tag{4}$$

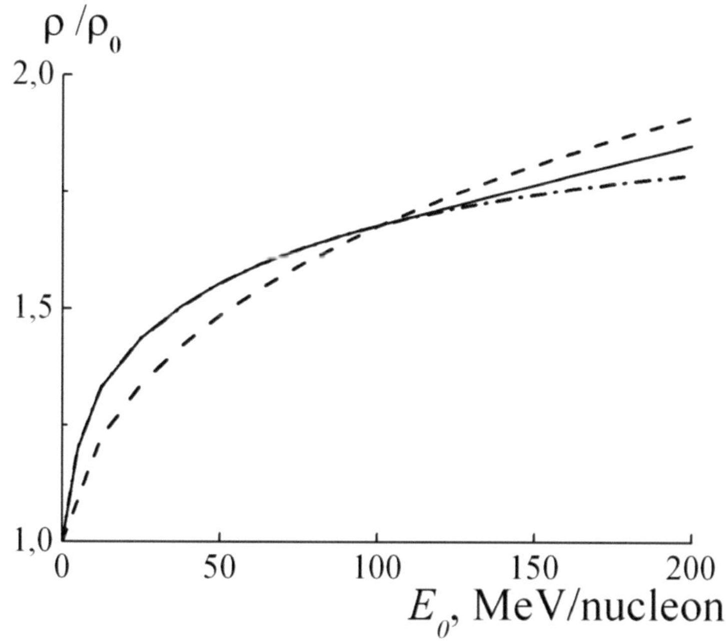

Figure 1.
The dependence of the maximum compression ratio ρ/ρ₀ on the collision energy E₀ achieved during the interaction of the overlapping regions of colliding nuclei for the case of the relaxation factor q calculated by us (solid line), for the case where the factor q = 0 (dashed line) and for the case where q = 1 (dash-dotted line). All dependences correspond to the value of the compression modulus K = 210 MeV.

$$\frac{\partial(m\rho v^2/2+e)}{\partial t}+\frac{\partial(r^2 v(m\rho v^2/2+e+P))}{r^2\partial r}=0. \tag{5}$$

The heat flux for a local equilibrium distribution function is $Q=0$. Here, the internal energy density is $e=e_{kin}+e_{int}$ and pressure is $P=P_{kin}+P_{int}$, where e_{kin} and P_{kin} are the kinetic terms, and the interaction terms e_{int} and P_{int} are.

$$e_{int}=\int_0^\rho W(\rho)d\rho, P_{int}=\rho^2\frac{d(e_{int}/\rho)}{d\rho}. \tag{6}$$

The velocity field is found from Eq. (3) in the approximation of a homogeneous but time-dependent density of hot spot $\rho(r,t)=\rho(t)$:

$$v(r,t)=\frac{\dot{R}_1}{R_1}r, 0\le r\le R_1, \tag{7}$$

$$v(r,t)=\frac{\dot{R}(r-R_1)-\dot{R}_1(r-R)}{(R-R_1)}, R_1\le r\le R, \tag{8}$$

where $R(t)$ is the radius of the hot spot; $R_1(t)$ is the radius of the velocity field kink determined from the solution of equations; and $\dot{R}(t)$ and $\dot{R}_1(t)$ are the derivatives in time (speed), which are also found from the Eqs. A system of ordinary integro-differential equations is obtained after integrating Eqs. (4) and (5) over the hotspot volume. It is solved numerically.

However, the deviation of the distribution function $f\left(\vec{r},\vec{p},t\right)$ from the local equilibrium function $f_0\left(\vec{r},\vec{p},t\right)$ is not taken into account in these equations. Expressing $f\left(\vec{r},\vec{p},t\right)$ from the right side of Eq. (1) through its left side, we find

$$f = f_0 - \tau\left(3 - \frac{5}{3}I\frac{\partial}{\partial I}\right)f_0\frac{\partial \upsilon}{\partial r}, \tag{9}$$

where I is the thermal term depending on the temperature T. When obtaining (9), we substituted $f_0\left(\rho(r,t), U\left(r,\vec{p},t\right), T(r,t)\right)$ into the left part of Eq. (1) instead of $f\left(r,\vec{p},t\right)$, taking into account Eqs. (3)–(5), where $U = \frac{\left(\vec{p}-m\vec{\upsilon}\right)^2}{2m}$. In this case, the hot spot was averaged over the volume to derive Eq. (1), and at the expansion stage, the density $\rho(r,t)$, the temperature $T(r,t)$, and the thermal term $I(r,t)$ were considered to be homogeneous functions of time t and independent of the radius r. Substituting expression (9) into the equations of hydrodynamics [11, 12], we find the corrections to kinetic terms of the energy density e_{kin} and pressure P_{kin}:

$$e_{kin} = e_{0,kin} - \tau\frac{4}{3}\left(e_{0,kin} + \frac{5}{4}e_F\right)\frac{\partial \upsilon}{\partial r} = e_{0,kin} - \frac{3}{2}\eta\frac{\partial \upsilon}{\partial r}, \tag{10}$$

$$P_{kin} = P_{0,kin} - \tau\frac{4}{3}\left(P_{0,kin} + \frac{5}{6}e_F\right)\frac{\partial \upsilon}{\partial r} = P_{0,kin} - \eta\frac{\partial \upsilon}{\partial r}. \tag{11}$$

where $e_{0,kin} = e_F + I$ and $P_{0,kin} = \frac{2}{3}e_{0,kin}$ are the equilibrium kinetic parts of the energy density and pressure density, $e_F = \frac{3}{10}\frac{\hbar^2}{m}\left(\frac{3}{2}\pi^2\rho\right)^{2/3}\rho$, and $\eta = \frac{4}{3}\left(P_{0,kin} + \frac{5}{6}e_F\right)\tau$ is the viscosity coefficient. The following correction terms turn out to be an order of magnitude smaller and they are not taken into account. The heat flux is $Q = 0$. The corrections to kinetic terms significantly affect the hotspot expansion and slow it down, because the Reynolds number is not large $Re = \frac{m\rho\upsilon l}{\eta} \sim 1$ for the viscosity coefficient η found by us (formula (10)) in the energy range under consideration of $E_0 \approx 100$ MeV per nucleon with a characteristic nuclear size of $l \approx 3$ fm. In our case, the temperature is $T \approx 20$ MeV; $P_{0,kin} \approx \rho T$; $\tau \approx 3 \cdot 10^{-23}$ s; the viscosity coefficient is $\eta \approx 4 \cdot 10^{10}$ kg m^{-1} s^{-1}. It coincides in the order of magnitude with the gas estimate $\eta \sim \sqrt{mT}/\sigma$ [16] if we take $\sigma \approx 40$ mb for the elementary cross section. Moreover, $\eta s >> \frac{\hbar}{4\pi}$, where s is the entropy density ($s \sim \rho$). That is, in our case, the ratio $\frac{\eta}{s}$ is more than an order of magnitude higher than the limiting value of $\frac{\hbar}{4\pi}$ [17] (achievable, e.g., in the state of a quark-gluon plasma). Thus, the viscosity coefficient is quite large in the energy range under consideration. This reduces the expansion speed of the hot spot and increases its temperature. Secondary particles (nucleons, fragments, and pions) form and freeze out when the expanding nuclear system reaches a critical density (freezing density) ρ^* determined from the condition $\frac{dP_{int}}{d\rho} = \rho\frac{dW}{d\rho} = 0$.

4. Double differential cross sections of the emission of protons and pions: comparison with the experimental data

Protons and pions are emitted when the nuclear system reaches a critical density. The cross section of the emission of protons (pions) is found from the condition that the number of particles $fd^3\vec{p}$ and the value $d^3\vec{p}/E$ of are relativistic invariants [18, 19]. As a result, the inclusive double differential cross section of reaction A + B → p(π) + X is

$$\frac{d^2\sigma}{dEd\Omega} = (2S+1)\frac{2\pi}{(2\pi\hbar)^3}\int G(b)bdb\int d\vec{r}\gamma\left(E-\vec{p}\vec{v}\right)pf\left(\vec{r},\vec{p},t\right), \tag{12}$$

where b is an impact parameter and the distribution function of protons (pions) has the form

$$f\left(\vec{r},\vec{p},t\right) = \left[\exp\left(\frac{\gamma\left(E \quad \vec{p}\vec{v} \quad \mu\right)}{T}\right) \pm 1\right]^{-1}. \tag{13}$$

Here E and $\vec{p}$ are the total energy and momentum of the proton (pion), respectively; $E = \sqrt{p^2 + m^2_{p(\pi)}}$; Ω is the solid angle; S is the spin; $\vec{v}\left(\vec{r},t\right)$ and $T\left(\vec{r},t\right)$ are the velocity field and temperature at the time of freeze-out (they are solutions of the equations of hydrodynamics); $\gamma = 1/\sqrt{1-(v/c)^2}$ is the Lorentz factor; μ is the chemical potential (for pions $\mu = 0$, because the number of pions is not specified). The factor $G(b) = \sigma_t(b)/\sigma_g(b)$ introduced in (12) takes into account the difference between the total cross section and the geometric cross section, where $\sigma_t(b)$ is defined as the cross section of the formation of a hot spot for a given impact parameter b from two overlapping regions in colliding nuclei, and $\sigma_g(b)$ is equal to the geometric cross section of these overlapping regions. Here, the total cross section is always greater than geometric one, as in the case of the fusion of two nuclei comparable in size. In addition, the function $f\left(\vec{r},\vec{p},t\right)$ included in Eq. (12) was modified in comparison with Eq. (13) according to relation (2): the sign "+" refers to protons, and the sign "−" refers to pions. Expressions (12) and (13) refer to protons (pions) emitted from a hot spot as a result of the interaction of the overlapping regions of colliding nuclei. In addition to this contribution, we took into account the contribution from the emission of protons (pions) as a result of the fusion of non-overlapping regions of colliding nuclei. The calculated double differential cross sections of proton emission (energy spectra) were compared with similar calculations obtained by solving the Vlasov-Uling-Uhlenbeck (VUU) kinetic equation [1] and with available experimental data. Our calculations corresponded to the equation of state with selected compression modulus equal to K = 210 MeV, i.e., with the same which was taken for the best description of the experiment in the calculations that we performed in [8, 9] at energies of 250 and 400 MeV per nucleon for colliding Ne and U nuclei.

We present the proton spectra in the ^{40}Ar + ^{40}Ca → p + X reaction at the angles of 30° (*1*), 50° (*2*), 70° (*3*), and 90° (*4*) for the energy of projectile nuclei of ^{40}Ar of 92 MeV per nucleon (**Figure 2**). In **Figure 2**, the solid curves correspond to our calculation, the histograms correspond to the calculations performed by the method of solving the VUU equation [1], and the dots are the experimental data from [1].

As can be seen, our calculation (this is not the Monte Carlo method and not histograms) is in good agreement with the experimental data. This is especially true for small angles of emission of protons (30°, 50°, and 70°). Our approach has an advantage over the more detailed method of solving the VUU equation [1], since the solid curves (but not histograms) are the result of the calculation. Note here that simple cascade models, as mentioned in [1], cannot describe these experimental data at all.

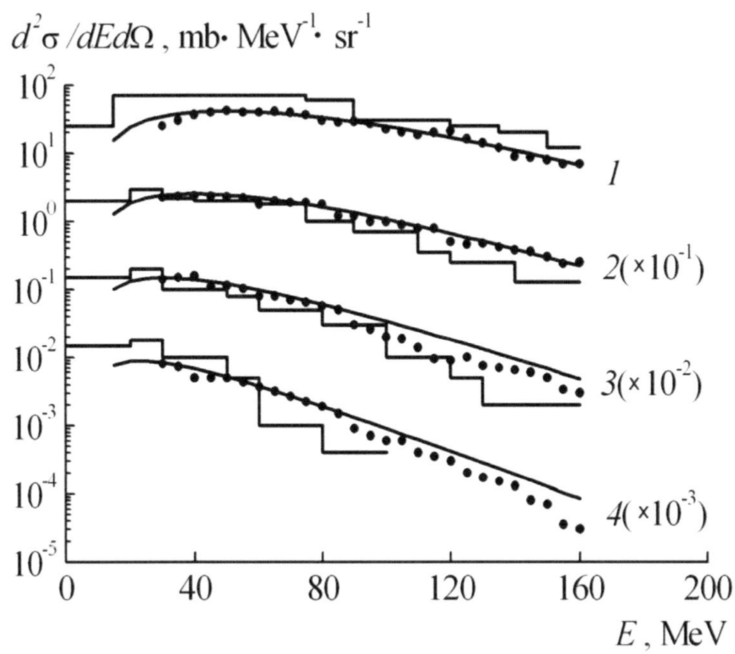

Figure 2.
Spectra of protons emitted in the reaction $^{40}Ar + {}^{40}Ca$ with the energy of ^{40}Ar ions of 92 MeV per nucleon at angles of 30° (1), 50° (2), 70° (3), and 90° (4). The solid lines are the results of calculations according to this model with the calculated q corresponding to K = 210 MeV; the histograms are the results of calculations obtained from the solution of the VUU kinetic equation (1); the dots are the experimental data from [1].

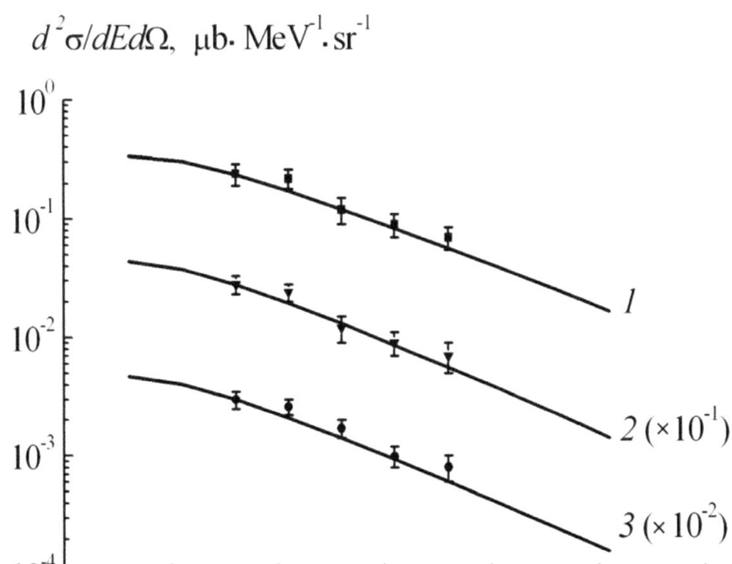

Figure 3.
The calculated (solid lines) and experimental (dots) [14] inclusive double differential cross sections of the emission of mesons at the observation angle of 90° in the reactions $^{16}O + {}^{27}Al$ (1), $^{16}O + {}^{58}Ni$ (2), and $^{16}O + {}^{197}Au$ (3) with energy of ^{16}O ions of E_0 = 94 MeV per nucleon.

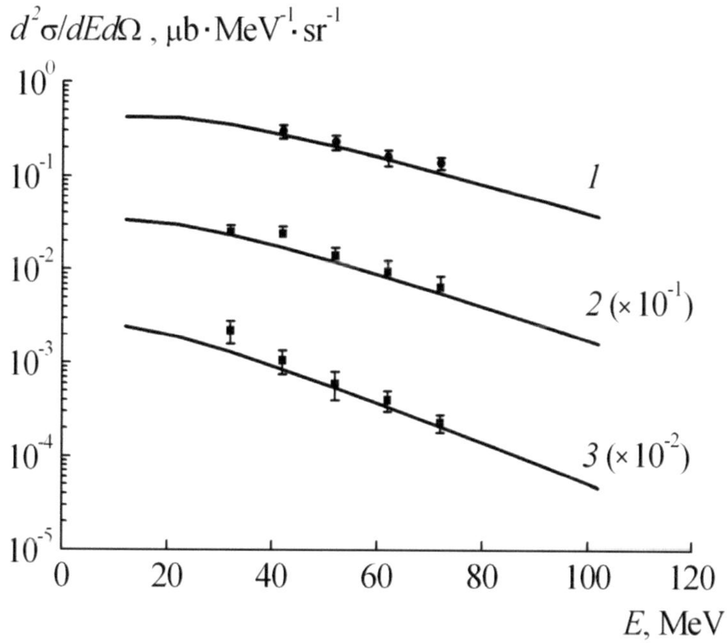

Figure 4.
The calculated (solid curves) and experimental (dots) [14] inclusive double differential cross sections of the emission of mesons in the reaction ^{16}O + ^{27}Al with energy of ^{16}O ions of 94 MeV per nucleon at the observation angles of 70° (curve 1), 90° (curve 2), and 120° (curve 3).

We compared our data with the available experimental data on the emission of pions. **Figure 3** illustrates the comparison of our calculated (solid lines) and experimental [14] (dots) double differential cross sections for the reactions of π^+-meson production when ^{16}O ions collide with ^{27}Al nuclei (curve *1*), ^{58}Ni nuclei (curve *2*), and ^{197}Au nuclei (curve *3*) at energies of ^{16}O ions of $E_0 = 94$ MeV per nucleon at an angle of 90°. It can be seen that the calculation is in good agreement with the experiment for chosen parameters of the nuclear interaction and taking into account the viscosity of the medium η that is found by us and proportional to the relaxation time τ within the experimental errors. In this case, the effect of viscosity on the calculated cross section of emitted pions is stronger for more asymmetric combinations of colliding nuclei, when the contribution of the emission of pions from the hot spot prevails. Thus, inclusive pion spectra in asymmetric nuclear collisions can be used to measure the viscosity of a nuclear medium.

Figure 4 illustrates the comparison of the calculations (solid lines) with the experimental data [14] (dots) for the reaction $^{16}O + {}^{27}Al \rightarrow \pi^+ + X$ at energy of ^{16}O ions of 94 MeV per nucleon at pion emission angles of 70° (curve *1*), 90° (curve *2*), and 120° (curve *3*). The calculation is in agreement with the experimental data if its parameters are constant. In all the illustrations under consideration, the agreement of calculation with the experiment was achieved without introducing fitting parameters and is more successful than our previous works [11, 19, 20].

5. Conclusions

Thus, the idea of using the hydrodynamic approach with a non-equilibrium equation of state in describing collisions of heavy ions is further developed in this work. The non-equilibrium equation of state is found in the approximation of the functional on the local density. The differential cross sections of the emission of protons and the production of subthreshold pions in heavy ion collisions are uniformly described with the same fixed parameters of the equation of state and in the same approach as in the previous papers [11–13], which describe the differential cross sections for the formation of protons and light fragments. It is shown that the non-equilibrium equation of state included in the hydrodynamic equations allows us to describe the experimental energy spectra of protons produced in collisions of heavy ions with intermediate energies better than the equation of state corresponding to traditional hydrodynamics, which initially implies the local thermodynamic equilibrium.

This simplified hydrodynamic approach including a description of the stages of compression, expansion, and freeze-out of a substance during heavy ion collisions turned out to be no worse than a more detailed approach based on the Monte Carlo solution of the Vlasov-Uling-Uhlenbeck kinetic equation.

In comparison with previous works, the inclusion of the effects of nuclear viscosity, which we found in the relaxation approximation for the kinetic equation, is new. This did not add new parameters in describing the temporal evolution of nuclear collisions. The relaxation time τ, which determines the nuclear viscosity coefficient η, turned out to be close to the value found on the basis of the behavior of nuclear Fermi liquid [15] and is not a fitting parameter. When describing the emission of protons and fragments, the inclusion of the viscosity of the medium is not so significant, and the pions are very sensitive to the viscosity.

The highlighting of proton (pion) emission after the temporal evolution of the resulting hot spot and the contribution to the particle emission cross sections during the fusion of "spectators" (non-overlapping regions of colliding nuclei) were significant in calculating the cross sections. This made it possible to describe the differential cross sections of the emission of protons (pions) for collisions of nuclei in various combinations. Highlighting this feature of our approach can be useful in comparison with other ways of pion production in heavy ion collisions, for example [21, 22], based on the solution of the Vlasov-Uling-Uhlenbek equation. These works include a range of higher energies of colliding heavy ions (more than 300 MeV per nucleon) and the production of pions by means of Δ-isobar production. We included this channel at low subthreshold energies, not limited to the production of thermal pions. However, this channel appears on the higher energy tails of the energy spectra of pions [23].

Studies of the formation of protons, fragments, and subthreshold production of pions may be of interest for the development of a scientific program planned with radioactive beams in Dubna using the COMBAS facility [24], which is designed to study nuclear collisions in the energy range of 20–100 MeV per nucleon.

Author details

A.T. D'yachenko[1]* and I.A. Mitropolsky[2]

1 Emperor Alexander I St. Petersburg State Transport University, St. Petersburg, Russia

2 Petersburg Nuclear Physics Institute, National Research Center Kurchatov Institute, Gatchina, Russia

*Address all correspondence to: dyachenko_a@mail.ru

References

[1] Stocker H, Greiner W. High energy heavy ion collisions—probing the equation of state of highly excited hardronic matter. Physics Reports [Internet]. Elsevier BV; May 1986;**137** (5-6):277-392. DOI: 10.1016/0370-1573 (86)90131-6

[2] Khvorostukhin AS, Toneev VD. Hadron rapidity spectra within a hybrid model. Physics of Particles and Nuclei Letters [Internet]. Pleiades Publishing Ltd; January 2017;**14**(1):9-17. DOI: 10.1134/s1547477117010162

[3] Khvorostukhin AS, Toneev VD. Rapidity distributions of hadrons in the HydHSD hybrid model. Physics of Atomic Nuclei [Internet]. Pleiades Publishing Ltd; March 2017;**80**(2):285-298. DOI: 10.1134/s1063778817010112

[4] D'yachenko AT. Physics of Atomic Nuclei. 1994;**57**:1930

[5] D'yachenko AT, Gridnev KA. Calculation scheme of heavy-ion collisions within the framework of a modified hydrodinamic approach. Bulletin of the Russian Academy of Sciences: Physics [Internet]. Allerton Press; July 2013;**77**(7):857-861. DOI: 10.3103/s1062873813070095

[6] D'yachenko AT, Gridnev KA, Greiner W. Calculation of heavy ion collisions within the framework of the modified hydrodynamic approach. Journal of Physics G: Nuclear and Particle Physics [Internet]. IOP Publishing; 20 June 2013;**40**(8): 085101. DOI: 10.1088/0954-3899/40/8/085101

[7] D'yachenko AT, Gridnev KA. Calculation of heavy-ion collisions within the framework of the hydrodynamic approach with a nonequilibrium equation of state. Bulletin of the Russian Academy of Sciences: Physics [Internet]. Allerton Press; July 2014;**78**(7):648-653. DOI: 10.3103/s1062873814070107

[8] D'yachenko AT, Gridnev KA, Mitropolsky IA. Features of a nonequilibrium equation of state in heavy-ion collisions at intermediate energies. Bulletin of the Russian Academy of Sciences: Physics [Internet]. Allerton Press; July 2015;**79**(7):858-863. DOI: 10.3103/s1062873815070096

[9] AT D'yachenko, K. A. Gridnev, I. A. Mitropolsky, and W. Greiner. In: YE Penionzhkevich and YG Sobolev editors. Proceedings of the 7th International Symposium on Exotic Nuclei, Kaliningrad, Russia, 2014, World Scientific, Singapore; 2015. p. 413

[10] D'yachenko AT, Mitropolsky IA. Vopr. At Nauki Tekh., Ser.: Yad.-Reakt. Konst. 2016;**2**:94

[11] D'yachenko AT, Mitropolsky IA. A nonequilibrium equation of state in heavy-ion collisions at intermediate energies. Bulletin of the Russian Academy of Sciences: Physics [Internet]. Allerton Press; August 2016; **80**(8):916-920. DOI: 10.3103/s1062873816080104

[12] D'yachenko AT, Mitropolsky IA. Fragments in heavy ion collisions in a hydrodynamic approach with a nonequilibrium equation of state. Bulletin of the Russian Academy of Sciences: Physics [Internet]. Allerton Press; December 2017;**81**(12):1521-1526. DOI: 10.3103/s1062873817100112

[13] D'yachenko AT, Mitropolsky IA, Sobolev YG. In: Penionzhkevich YE, Sobolev YG, editors. Proceedings of the 8th International Symposium on Exotic Nuclei, Kazan, Russia, 2016, World Scientific, Singapore; 2017. p. 38

[14] Badala A, Barbera R, Bonasera A, Di Toro M, Palmeri A, Pappalardo GS, et al.

Statistical and microscopic description of energetic products in the reactions induced by O16 on Al27, Ni58, and Au197 at 94 MeV/nucleon. Physical Review C [Internet]. American Physical Society (APS); 1 January 1991;**43**(1): 190-200. DOI: 10.1103/physrevc.43.190

[15] Bertsch G. The collision integral in nuclear matter at zero temperature. Zeitschrift für Physik A. Atoms and Nuclei [Internet]. Springer Science and Business Media LLC; March 1978;**289** (1):103-115. DOI: 10.1007/bf01408501

[16] Lifshitz EM, Pitaevskii LP. Course of Theoretical Physics. In: Physical Kinetics. Vol. 10. (Pergamon, Oxford, 1981; Nauka, Moscow, 1979). [Publishing: 1 January 1981]

[17] Kovtun PK, Son DT, Starinets AO. Viscosity in strongly interacting quantum field theories from black hole physics. Physical Review Letters [Internet]. American Physical Society (APS); 22 March 2005;**94**(11). DOI: 10.1103/physrevlett.94.111601

[18] Das Gupta S, Mekjian AZ. The thermodynamic model for relativistic heavy ion collisions. Physics Reports [Internet]. Elsevier BV; June 1981;**72**(3): 131-183. DOI: 10.1016/0370-1573(81) 90012-0

[19] D'yachenko AT. Bulletin of the Russian Academy of Sciences: Physics. 1998;**62**:170

[20] D'yachenko AT, Lozhkin OV. Nuclear Physics. 1997;**A696**:81

[21] Zhang Z, Ko CM. Medium effects on pion production in heavy ion collisions. Physical Review C [Internet]. American Physical Society (APS); 7 June 2017;**95** (6). DOI: 10.1103/physrevc.95.064604

[22] Cheng S-J, Yong G-C, Wen D-H. Effects of the symmetry energy in the Sn132+Sn124 reaction at 300 MeV/ nucleon. Physical Review C [Internet]. American Physical Society (APS); 30 December 2016;**94**(6). DOI: 10.1103/ physrevc.94.064621

[23] D'yachenko AT, Baryshnikov VN. On the role of Δ-resonance in the pion production in collisions of heavy intermediate-energy ions in the hydrodynamic approach. Bulletin of the Russian Academy of Sciences: Physics [Internet]. Allerton Press; June 2009;**73** (6):724-726. DOI: 10.3103/ s1062873809060069

[24] Artukh AG, Klygin SA, Kononenko GA, Kyslukha DA, Lukyanov SM, Mikhailova TI, et al. Radioactive nuclear beams of COMBAS facility. Physics of Particles and Nuclei [Internet]. Pleiades Publishing Ltd; January 2016;**47**(1):49-72. DOI: 10.1134/ s1063779616010032